Intervention
Teaching Guide

SAXON MATH™
5/4

Pat Wrigley

SAXON™
PUBLISHERS

Saxon Publishers gratefully acknowledges the contributions of the following individuals in the completion of this project:

Author: Pat Wrigley

Editorial: Brian E. Rice, Andrew Kershen

Editorial Support Services: Chris Davey, Jenifer Sparks, Shelley Turner, Stephen Gill, Cassandra Phillips, Jean Van Vleck, Darlene Terry

Production: Tara Robinett, Julie Webb, Cristi D. Whiddon

Project Management: Angela Johnson, Becky Cavnar

Saxon Publishers would especially like to thank the following teachers, who provided invaluable feedback in the development of this product:

Latrelle McDonald, Huntsville, North Carolina; Jeanne Nunn, Lenexa, Kansas; Carol Rupert, Lenexa, Kansas; Margaret Wagner, Muskogee, Oklahoma

Printed in the United States of America

ISBN: 1-59141-277-3

Manufacturing Code: 1 2 3 4 5 6 7 8 022 12 11 10 09 08 07 06 05

From the Author

Forty years of teaching has led me to a few discoveries that are worth sharing. And as a recovering math-phobe for the past eighteen years, I find I have become positively evangelical about communicating my joy at having discovered Saxon Math™.

My first twelve years in the classroom were spent teaching Grades 4 through 8 in general education, in both public and private schools. I later received my Master's Degree in Special Education of Physically Handicapped Students. Since then, most of my teaching experience has been as a resource specialist in the middle school or junior high setting.

I was effective at teaching language arts, social studies, and foreign languages, and even reasonably good in science; but before I discovered Saxon, I had become very discouraged teaching math. My students could learn the material, but they couldn't retain it. So I spent two years writing my own textbook, stressing mastery plus maintenance of skills learned. Then, at a math conference one summer, I discovered that Stephen Hake and John Saxon of Saxon Publishers had not only arrived before me, but had surpassed my finest efforts.

In the past eighteen years of teaching with Saxon books, I have learned to love teaching math. Without hesitation I can say it is my favorite subject. Every teacher should get such satisfaction from his or her work! In unlocking the mysteries of math I get to give my kids a running jump at self-esteem. My students know they are terrific and show it. I always was a competent teacher, but until I started using Saxon, I never touched the lives of my students so effectively.

I have learned a great deal since my first year of using Saxon, and you will too. If the materials I have developed overwhelm you at first, start with only what you can comfortably use. You will grow with the program as I did. I hope this teaching guide is helpful. It will be a resource to refer to as you refine your approaches. And please share new approaches with me. I'm always fine-tuning my program.

Acknowledgments

I would like to recognize the people who saw value in my work and encouraged it from the beginning. They are Frank Wang and Deborah Hicks of Saxon Publishers, and my principal at Vallejo High, Phil Saroyan. Thanks, too, to the editors who polished my raw material: Julie Webster, Brian Rice, Mary Burleson, Sean Douglas, Clint Keele, Andrew Kershen, Sherri Little, Marta Matsumoto, and Brian Smith. Finally, a special thanks to my daughter Sarah, whose special needs inspired my work.

Pat Wrigley
Resource Specialist
Vallejo High School
Vallejo, California

Contents

Overview

Intervention *materials provide alternate teaching strategies, assistance for the lessons and tests, student reference materials, remediation support, and extra practice.*

Saxon Math 5/4 Intervention provides support materials for the instruction and exercises in the *Saxon Math 5/4* program to assist struggling learners. The *Intervention* materials are designed to ensure success for a range of students, including students who have difficulties with the following:

- visual-motor integration
- fine motor coordination
- spatial organization
- receptive language
- number transpositions in copy work
- getting started
- verbal explanations
- distractibility
- math anxiety

The *Intervention* materials provide alternate teaching strategies, assistance for the lessons and tests, student reference materials, remediation support, and extra practice on select topics. They can be used in a variety of classroom settings, including inclusion and pullout programs.

This guide explains how to take advantage of the *Intervention* materials according to the needs of your classroom. It offers support for minimal to maximal implementation depending on the instructional model, teacher resources, students' abilities, and classroom materials.

Components

The *Intervention* materials described below are designed to support the *Saxon Math 5/4* program, so they must be used in conjunction with the Student Edition, Teacher's Manual, and Assessments and Classroom Masters.

Teaching Guide

This *Saxon Math 5/4 Intervention* Teaching Guide explains how to assist struggling students in their study of *Saxon Math 5/4*. In addition to providing procedural information, it contains fifty-five teaching hints that discuss strategies for communicating potentially difficult concepts to students. The teaching hints are referenced in the *Saxon Math 5/4* Teacher's Manual so they can be used in planning instruction.

Student Reference Guide

Daily use of the reference guide promotes memory of important math concepts.

Many students have difficulty memorizing facts such as weight equivalences and characteristics of geometric figures. To ease the burden of recalling information, the eight-page *Saxon Math 5/4–6/5 Intervention* Student Reference Guide includes charts, formulas, conversions, techniques, definitions, and other mathematical facts. As students grow familiar with the guide, they will gain confidence and become more proficient math students.

Reference guides may be copied. They are packaged with the Teacher's Manuals and with the Student Workbooks (see below); they are also available for separate purchase in sets of eight. Provide one for each struggling student, and emphasize its value as a resource for solving problems.

Student Workbook

The *Saxon Math 5/4 Intervention* Student Workbook contains a worksheet for every lesson, investigation, and supplemental practice in the *Saxon Math 5/4* textbook. Individual copies of the workbook may be purchased for each student, or you may use a single workbook as a book of blackline masters.

The worksheets provide teacher notes, easy-to-understand lesson summaries, and adequate, formatted workspace for completing problems and recording answers. (See the annotated sample on page 4.) To reward completion of a grade level, a graduation certificate appears at the end of the workbook.

Masters

The *Saxon Math 5/4 Intervention* Masters contains additional blackline masters designed to assist struggling students. It includes Test Worksheets, which adapt the regular tests by providing clues and workspace for the problems; Test-Item Analyses for remediation purposes; Answer Forms to facilitate the transition from worksheet to notebook-paper use; Recording Forms for tracking and analyzing performance; Sequence Sheets, which identify the order of assignments; Reference Charts that can be enlarged and posted in the classroom; and *Intervention*-specific parent letters.

Answer Key

Intervention materials often explain alternative methods for solving problems. Thus, they require a separate answer key. The *Saxon Math 5/4 Intervention* Answer Key contains answers to all *Saxon Math 5/4 Intervention* materials as well as full solutions to all Test Worksheets. For convenience, answers to masters from the Assessments and Classroom Masters (specifically, the Facts Practice Tests, Test-Day Activity Masters, and Activity Masters) are also included.

Preparing Your Classroom

Preparation tips:
- *Copy one week's worth of assignments at a time.*
- *Store manipulatives where easily accessible.*
- *Display Reference Charts and Posters on the days you teach related lessons.*

A well-organized, smoothly running classroom enhances instruction. Students should know where to turn in and pick up assignments, where to find manipulatives, and when it is appropriate to do these things.

Each struggling math student should receive a copy of the Student Reference Guide at the beginning of the year. If you copy each page onto a different color of paper, students will have a visual reminder of where information appears in the reference guide. To make the pages more durable, consider lamination. Students might also wish to keep a blank piece of paper with their reference guides to record additional definitions and formulas as needed.

The amount of support material you will need to copy depends on whether students have individual copies of the Student Workbook. If they do, you will need to make copies of support materials from the *Intervention* Masters every few days. More copying will be required for teachers who use the Student Workbook as a book of masters. Those teachers may wish to copy a week's worth of Lesson Worksheets at a time for struggling students. As student usage fades throughout the year (see "Decreasing Usage of Support Materials" on page 59), the number of copies required each week will decrease. The worksheets can be filed in hanging folders or three-hole punched and kept in a binder. Consider storing the worksheets in an easily accessible location so that students can find them independently.

For many tactile and kinesthetic learners, use of manipulatives is indispensable. Consider establishing a station where students can access them with ease. The *Saxon Math 5/4–8/7* Manipulative Kit can be purchased to supplement the program, or you can stock the station with any manipulatives you already have.

For visual learners, post the Reference Charts from the *Intervention* Masters, as well as the Concept Posters that are available for optional purchase. The Reference Charts may be enlarged for easier viewing. It is not necessary to post all the charts at the beginning of the year, but you might want to arrange the classroom so that there is a convenient place to post them at the opportune time. A schedule in the *Intervention* Masters indicates the ideal lesson with which to post each chart.

Adapting for Struggling Students

Meeting the needs of struggling math students is challenging, but the *Saxon Math 5/4 Intervention* materials help teachers achieve that goal. The worksheets are key. Below are descriptions of the worksheets, followed by information about how to adapt your classroom routine to meet students' special needs. This guide explains a range of ideas because variations in teaching models, teaching resources, student learning abilities, and classroom makeup present unique opportunities and challenges. As you read the following pages, be mindful of your situation; not every idea will be necessary or appropriate for your class.

Worksheets

Worksheets must be used in conjunction with the Student Edition.

The worksheets in the *Saxon Math 5/4 Intervention* Student Workbook provide support for the instruction and daily practice problems in the *Saxon Math 5/4* Student Edition. Each Lesson Worksheet summarizes the lesson by highlighting key facts, examples, and procedures. Because the lesson summaries are brief, students are more likely to read them. Reliance on simpler language ensures that more students will understand and remember the concepts.

The worksheets' greatest value is the support they provide for practice problems, including:

- identifying a starting point
- restating the problem
- crafting a partial solution
- citing a Student Reference Guide page
- referring students to a page in the textbook
- reminding students to include units in the answer

The level of support for individual problems depends on many factors, including the complexity of the topic, the frequency of related problems, and how recently the topic was introduced. As procedures and ideas become more familiar, support gradually decreases. For some topics it fades completely, but for especially difficult topics it does not.

The annotated sample on the following page highlights many features and benefits of the worksheets.

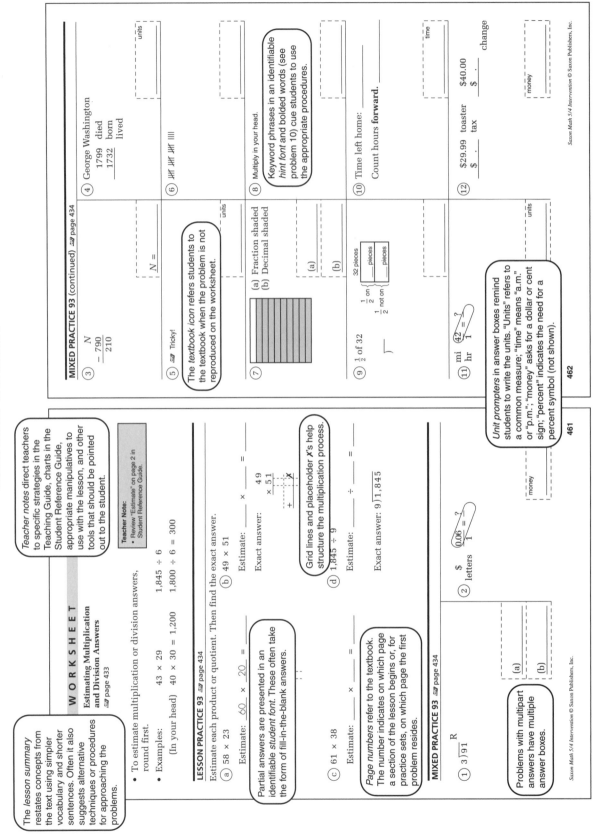

Annotated Worksheet†

The *lesson summary* restates concepts from the text using simpler vocabulary and shorter sentences. Often it also suggests alternative techniques or procedures for approaching the problems.

Teacher notes direct teachers to specific strategies in the Teaching Guide, charts in the Student Reference Guide, appropriate manipulatives to use with the lesson, and other tools that should be pointed out to the student.

WORKSHEET

Estimating Multiplication and Division Answers
page 433

Teacher Note:
• Review "Estimate" on page 2 in Student Reference Guide.

• To estimate multiplication or division answers, round first.
• Examples: 43×29 $1,845 \div 6$
  (In your head) $40 \times 30 = 1,200$ $1,800 \div 6 = 300$

LESSON PRACTICE 93 page 434
Estimate each product or quotient. Then find the exact answer.

(a) 58×23
Estimate: _60_ × _20_ = ___

(b) 49×51
Estimate: ___ × ___ = ___
Exact answer:
$$\begin{array}{r} 4\,9 \\ \times\ 5\,1 \\ \hline + \end{array}$$

Partial answers are presented in an identifiable *student font*. These often take the form of fill-in-the-blank answers.

(c) 61×38
Estimate: ___ × ___ = ___

(d) $1,845 \div 9$
Estimate: ___ ÷ ___ = ___
Exact answer: $9\overline{)1,845}$

Grid lines and placeholder *x*'s help structure the multiplication process.

MIXED PRACTICE 93 page 434

(1) $3\overline{)91}$ R

(2) letters $ [0.06 / 1 = ?]
(a)
(b)

Problems with multipart answers have multiple answer boxes.

Page numbers refer to the textbook. The number indicates on which page a section of the lesson begins or, for practice sets, on which page the first problem resides.

money

461

MIXED PRACTICE 93 (continued) page 434

(3) N
   $-\ 790$
   210
$N = $ ___

(4) George Washington
   1799 died
   1732 born
   ___ lived
units

(5) Tricky!

The *textbook icon* refers students to the textbook when the problem is not reproduced on the worksheet.

(6) 卌 卌 卌 卌 III
$N = $? units

(7) [shaded figure]
(a) Fraction shaded
(b) Decimal shaded

Multiply in your head.

(8) [hint]

Keyword phrases in an identifiable *hint font* and bolded words (see problem 10) cue students to use the appropriate procedures.

(9) $\frac{1}{2}$ of 32
32 pieces $\frac{1}{2}$ on ___ pieces $\frac{1}{2}$ not on ___ pieces
(a)
(b)

(10) Time left home: ___ time
Count hours **forward**.

(11) mi [42 / 1 = ?]
     hr
units

(12) $29.99 toaster
     $. tax
     $40.00
     $. change
money

Unit prompters in answer boxes remind students to write the units. "Units" refers to a common measure; "time" means "a.m." or "p.m."; "money" asks for a dollar or cent sign; "percent" indicates the need for a percent symbol (not shown).

462

†Lesson Worksheets are four pages long. Only two pages are shown here.

Lessons

Each lesson consists of four parts: Warm-Up, New Concept, Lesson Practice, and Mixed Practice. Students should focus on the practice segments for a majority of the math period. Keep the Warm-Up and New Concept sections brief.

Warm-Up

Daily warm-up exercises help focus students' attention on math.

The Warm-Up begins with a **Facts Practice Test** from the *Saxon Math 5/4 Assessments and Classroom Masters*. This short, timed exercise builds mastery of fundamental math knowledge that empowers student achievement. Learning basic facts is a significant and genuine accomplishment that can enhance a student's sense of success in and motivation for the subject. Administer the tests as directed in the *Saxon Math 5/4* Teacher's Manual, noting the following exceptions:

- Students with weak visual memory skills will have difficulty memorizing basic math facts. Facts Practice Tests sometimes help, but the gains can be disproportionate to the time and effort spent. This is especially true for students over the age of twelve. With such students, weigh the decision to use facts practice carefully.

- You may customize facts practice to match the needs of students. Memorizing basic facts can be challenging, even daunting for some. Acquiring mastery requires learning a fact or two per day and continually practicing learned facts. Strive to minimize the endeavor. For example, although there are 100 different multiplication problems from 0×0 through 9×9, the rules for multiplication by 0 and 1 are simple, and you need only teach thirty-six facts from 2×2 through 9×9.[†] Most students know some of those facts, further limiting the number you need to teach.

- You may provide alternatives to timed written tests when necessary. While timed tests can stir excitement and motivation for some students, they can also create excessive anxiety for others. In addition, some students' Individualized Education Plans disallow timed exams. For such students, you may remove the time restriction or allow them to study flash cards while others complete the facts practice.

Following the facts practice is the **Mental Math** section of the Warm-Up. The mental math problems should be completed orally as a whole-group activity. The difficulty of a problem may be adjusted up or down depending on the abilities of individual students. (For example, if a problem asks students to count up by fives from 20, lower-ability students might be allowed to begin at 5 instead, while advanced students could be asked to count down from 83.) When a series of related problems appears, help the class make connections and discover the pattern. Finally, when selecting a student to answer, let your understanding of who needs practice with individual problems guide you.

[†]In total, there are sixty-four problems to teach from 2×2 through 9×9, but many are practically identical (for example, 5×4 and 4×5), leaving only 36 fundamentally different facts to teach.

The Warm-Up culminates with **Problem Solving.**[†] Approach the exercise much like Mental Math (as an oral, whole-group activity). Let students share ideas and collaborate to derive the answer. Provide them with Answer Form A from the *Intervention* Masters—it lists many useful problem-solving techniques. If students are stumped by a problem, suggest that they try a different strategy from the list. Also, refer to the Teacher's Manual for instructional suggestions.

As mentioned before, keep these warm-up events brief. Try to complete all three in fifteen minutes or less. They are intended to energize students' minds for math and establish a tempo for the remaining portions of the lesson.

New Concept

Enhance instruction with the Student Reference Guide, Student Workbook, Teaching Guide, and manipulatives.

The New Concept portion of the lesson presents new instruction and should usually take five to ten minutes. Students are not required to fully grasp the concepts during this initial presentation. They will gain understanding over time as they revisit the concepts in future Mixed Practice sets (discussed on page 7–8). You can take the following steps to maximize the value of this learning experience for struggling students:

- Point out charts in the Student Reference Guide whenever they relate to the lesson.

- Allow students to use Lesson Worksheets.

- Whenever possible, use the teaching hints from this guide (beginning on page 14). They discuss alternate teaching strategies and help prevent stumbling blocks.

- Incorporate manipulatives and visuals into your instruction.

Your instructional model will factor into how you can help struggling students. If you apply the **whole-group instruction** model, where all students study the same lesson at the same time, use the following ideas to address the needs of students with difficulties:

Incorporate the examples on the Lesson Worksheets when applying whole-group instruction.

- While the textbook might provide several examples to demonstrate the new concept, only one or two of those will be shown on the Lesson Worksheet. Be especially mindful of those examples, and plan your instruction around them. Students will retain more if what they observe on the board or hear during the discussion matches the information on their papers.

- Avoid wordiness and complexity.

- After the whole-group instruction, provide individualized instruction as needed while students are working on the practice sets. Be aware of each student's learning style, and tailor your teaching to the individual. If multiple students need help, call on paraprofessionals (if available) to assist some students.

Self-paced instruction allows students to progress at individual rates. Some students may be able to make gains of more than one grade level in a single year.

If your teaching model involves **self-paced instruction,** which allows students to progress through their textbooks at individual rates, the following ideas will help you meet the needs of students with difficulties:

- Allow students to teach themselves using their textbooks and the lesson summaries on the companion Lesson Worksheets. Check understanding by reviewing students' work on the

---

[†]Problem Solving is sometimes labeled "Patterns" or "Vocabulary."

Lesson Practice (described below). Provide individualized instruction as needed.

- Provide each student with a Sequence Sheet, which indicates the sequence of assignments. This will allow the student to work more independently. (More detail is given on pages 10–11.)

- Have paraprofessionals or assistant teachers score papers and monitor student progress. Also, have them prepare in advance any materials students will need for their lessons.

- Do not let students sit idle as they wait their turn for instruction. Instead, encourage them to work on the easier problems from their practice sets until you can help them.

- Make students accountable for completing at least one lesson (or other assignment) per class, including the corresponding practice sets. Also, have them correct any mistakes made the previous day.

- If possible, encourage students to complete two lessons per class. Though challenging, it will greatly benefit students who perform significantly below level. Some may be able to make gains of more than one grade level in a single school year. (This will likely require sixty- to ninety-minute math class periods.)

Lesson Practice

The Lesson Practice usually contains five to ten problems focusing on the day's topics. The worksheet provides the full text of most problems and sets up many of the solutions, giving more help at first and gradually reducing assistance throughout.

If you use whole-group instruction, solve the first few problems in the Lesson Practice as a class; then have students complete the remaining problems on their own or in small groups. After going over the first few problems, you (and any teaching assistants you might have) can circulate and help individuals with the lesson as needed.

In self-paced classrooms, have students begin Lesson Practice on their own. Allow them to ask questions as needed. Circulate through the classroom to review problems, check understanding, and provide instruction as required.

Mixed Practice

Reserve most of the math period for Mixed Practice.

After students complete the Lesson Practice, they should move directly into the Mixed Practice and work on the problems individually. Ideally, students will spend most of the math period on Mixed Practice. Use the following suggestions to enhance the learning opportunity that the Mixed Practice provides:

- Help students learn how to be resourceful problem solvers by teaching them how to use the materials that will improve their understanding of concepts (the reference guide, posters, manipulatives, textbook). Ensure easy access to manipulatives and other resources.

- Point out the Lesson Reference Numbers shown in parentheses under the problem numbers in the textbook. Each one identifies a lesson that students can review if they have forgotten how to solve the corresponding problem. Rereading lessons or reviewing examples may refresh students' memories.

- Circulate through the classroom and provide one-on-one help as needed. If multiple students need help at the same time, instruct students to work on more familiar problems until you or a teaching assistant can help them.

Investigations

Teacher-directed Investigations follow every tenth lesson. The problems focus on several related topics and are interspersed with instruction. Often Investigations involve activities that require more preparation than typical lessons.

Some of the hints in this guide identify ways to simplify the instruction in Investigations. Also, the *Saxon Math 5/4 Intervention* Student Workbook provides Investigation Worksheets that summarize the instruction and provide support for select problems.

Tests

Testing tips:
- *Strongly discourage calculator use (except with physically disabled students).*
- *Encourage reference guide use.*
- *Test in class only.*

Tests are administered on a regular basis. Each test is cumulative, covering content through a certain lesson. Concepts are assessed only after they have been practiced in at least five Mixed Practices.

For test days, the *Saxon Math 5/4* Teacher's Manual recommends a Facts Practice Test and a Test-Day Activity in addition to the assessment. (The Test-Day Activities are brief instructional events that often require Test-Day Activity Masters from the *Saxon Math 5/4 Assessments and Classroom Masters.* Usually they can be completed in ten to fifteen minutes.) The cumulative nature of the tests, the testing schedule, and the additional activities can pose difficulties for some students. To ensure success, you can make a number of adjustments to the normal routine.

- Allow students to use Test Worksheets from the *Intervention* Masters. Much like Lesson Worksheets, they provide support for solving individual problems, including the full text of every problem, workspace, answer blanks, and hints for select problems.

- Encourage students to use their Student Reference Guides as a resource during the exam.

Use the delayed testing schedule (schedule B—see following page) with students who need extra time to practice.

- If recent performance suggests students will struggle with their next test, delay the exam to allow more practice time. You might want to wait as many as five additional lessons before giving the exam. (This option is shown in the testing schedule on the following page.)

- Distribute tests from a common folder or bundle to avoid stigmatizing students who are using the *Intervention* Test Worksheets or who are on an alternate testing schedule.

- Although Facts Practice Tests are helpful, they are not critical on test days. If time is a concern, skip the Facts Practice Test.

- As a last resort, you may also skip the Test-Day Activity; however, bear in mind that the activity might address a concept on your state or district end-of-year exam. In such a case, use the activity at the beginning of the next math period.

Testing Schedule | This schedule contains two options for the sequence of exams. Schedule A corresponds to the sequence outlined in the Teacher's Manual. Schedule B is an option that can be used to allow struggling students additional practice time in advance of their exams.

Testing Schedule

TEST TO BE ADMINISTERED	COVERS THROUGH	SCHEDULE A: GIVE AFTER	SCHEDULE B: GIVE AFTER	IF NECESSARY, RECYCLE[†] TO
Test 1	Lesson 5	Lesson 10	Lesson 15	Lesson 1
Test 2	Lesson 10	Lesson 15	Lesson 20	Lesson 6
Test 3	Lesson 15	Lesson 20	Lesson 25	Lesson 11
Test 4	Lesson 20	Lesson 25	Lesson 30	Lesson 16
Test 5	Lesson 25	Lesson 30	Lesson 35	Lesson 21
Test 6	Lesson 30	Lesson 35	Lesson 40	Lesson 26
Test 7	Lesson 35	Lesson 40	Lesson 45	Lesson 31
Test 8	Lesson 40	Lesson 45	Lesson 50	Lesson 36
Test 9	Lesson 45	Lesson 50	Lesson 55	Lesson 41
Test 10	Lesson 50	Lesson 55	Lesson 60	Lesson 46
Test 11	Lesson 55	Lesson 60	Lesson 65	Lesson 51
Test 12	Lesson 60	Lesson 65	Lesson 70	Lesson 56
Test 13	Lesson 65	Lesson 70	Lesson 75	Lesson 61
Test 14	Lesson 70	Lesson 75	Lesson 80	Lesson 66
Test 15	Lesson 75	Lesson 80	Lesson 85	Lesson 71
Test 16	Lesson 80	Lesson 85	Lesson 90	Lesson 76
Test 17	Lesson 85	Lesson 90	Lesson 95	Lesson 81
Test 18	Lesson 90	Lesson 95	Lesson 100	Lesson 86
Test 19	Lesson 95	Lesson 100	Lesson 105	Lesson 91
Test 20	Lesson 100	Lesson 105	Lesson 110	Lesson 96
Test 21	Lesson 105	Lesson 110	Lesson 115	Lesson 101
Test 22	Lesson 110	Lesson 115	Lesson 120	Lesson 106
Test 23	Lesson 115	Lesson 120	Inv. 12	Lesson 111

[†]Recycling is one option for helping a student who does poorly on a test. For more information on this and other options, see the Remediation section of this guide on pages 55–58.

Sequence Sheets

Sequence Sheets help students in self-paced classrooms plan for future assignments.

Two charts showing options for the sequence of assignments are provided in the *Intervention* Masters. (Miniature versions are shown on the following page.) The charts list all Lessons, Investigations, Supplemental Practices, Fraction Activities, and Tests. The only difference between them is the testing schedule (schedule A versus B, both of which are described on the previous page).

In classes that use self-paced instruction, give each student the sequence sheet corresponding to the intended testing schedule so that he or she can determine what the next assignment will be. The sequence sheet may also serve as a recording sheet—as students complete assignments, they can record the date in the appropriate box or fill in the box with pencil or ink.

The first column on each sheet lists the Lessons and Investigations, the second lists the Supplemental Practices and Fraction Activities, and the third lists the Tests. All students should be accountable for the first and third columns (the Lessons, Investigations, and Tests). Students who have difficulty learning or retaining concepts should complete assignments in the second column (Supplemental Practices and Fraction Activities) as well. The Supplemental Practices are contained in the Student Workbook, and the Fraction Activity Masters are found in the *Intervention* Masters.

An alternative to the Sequence Sheets is Recording Form F in the *Intervention* Masters. It contains most of the same information but presents it differently.

Saxon Math 5/4 *Sequence Sheets*

Sequence Sheet B

Sequence Sheet B reflects testing schedule B.

Lesson	Supplemental Practice	Test
1		
2		
3		
4		
5		
6		
7		
8		
9		
10		
Inv. 1		
11		
12		
13		
14		
15		1
16	16	
17	17	
18		
19		
20		2
Inv. 2		
21		
22	A†	
23		
24		
25		3
26	B	
27		
28		
29		
30	30	4
Inv. 3		
31		
32		
33		
34	34	
35		5
36		
37	37	
38		
39		
40		6
Inv. 4, Part 1		
Inv. 4, Part 2		
Inv. 4, Part 3		
41	41	
42		
43	43	
44		
45		7
46		
47		
48	48	
49		
Inv. 5	C	
50	50	8
51		
52	52	
53	53	
54		
55		9
56	D	
57		
58	58	
59		
60		10
Inv. 6	E	
61		
62		
63		
64	64	
65	65	11
66		
67	67	
68	68	
69		
70		12
Inv. 7		
71		
72		
73		
74		
75		13
76	76	
77		
78		
79		
80	80	14
Inv. 8		
81		
82		
83		
84		
85		15
86		
87		
88		
89	F	
90	90	16
Inv. 9		
91		
92		
93		
94		
95	G	17
96		
97		
98		
99		
100		18
Inv. 10		
101		
102		
103		
104	104	
105		19
106		
107	H, 107	
108		
109	I	
110	110	20
Inv. 11		
111		
112	J, 112	
113	113	
114	114	
115	115	21
116		
117		
118	118	
119	119	
120		22
Inv. 12		23
		Final

†Letters refer to the Fraction Activity Masters.

Sequence Sheet A

Sequence Sheet A reflects testing schedule A.

Lesson	Supplemental Practice	Test
1		
2		
3		
4		
5		
6		
7		
8		
9		
10		1
Inv. 1		
11		
12		
13		
14		
15		2
16	16	
17	17	
18		
19		
20		3
Inv. 2		
21		
22	A†	
23		
24		
25		4
26	B	
27		
28		
29		
30	30	5
Inv. 3		
31		
32		
33		
34	34	
35		6
36		
37	37	
38		
39		
40		7
Inv. 4, Part 1		
Inv. 4, Part 2		
Inv. 4, Part 3		
41	41	
42		
43	43	
44		
45		8
46		
47		
48	48	
49		
Inv. 5	C	
50	50	9
51		
52	52	
53	53	
54		
55		10
56	D	
57		
58	58	
59		
60		11
Inv. 6	E	
61		
62		
63		
64	64	
65	65	12
66		
67	67	
68	68	
69		
70		13
Inv. 7		
71		
72		
73		
74		
75		14
76	76	
77		
78		
79		
80	80	15
Inv. 8		
81		
82		
83		
84		
85		16
86		
87		
88		
89	F	
90	90	17
Inv. 9		
91		
92		
93		
94		
95	G	18
96		
97		
98		
99		
100		19
Inv. 10		
101		
102		
103		
104	104	
105		20
106		
107	H, 107	
108		
109	I	
110	110	21
Inv. 11		
111		
112	J, 112	
113	113	
114	114	
115	115	22
116		
117		
118	118	
119	119	
120		23
Inv. 12		Final

†Letters refer to the Fraction Activity Masters.

Teaching Strategies

*Struggling students need **brief, specific** directions. The following teaching strategies will help explain new concepts simply and easily.*

Struggling students need brief, specific directions with supporting visuals and manipulatives. The following pages contain detailed teaching strategies and techniques to use when introducing new concepts. These strategies, called *hints* in this guide, are identified by number and are referenced in the *Saxon Math 5/4* Teacher's Manual at point of use. The table below identifies the appropriate lesson, Investigation (Inv.), or appendix topic (Top.) in which to introduce each hint; the hints are discussed beginning on page 14.

Teaching Strategies for *Saxon Math 5/4 Intervention*

HINT	CONCEPT	INTRODUCE IN LESSON	TEACHING GUIDE PAGE
1	Column Addition (Sets of Ten)	1	14
2	Word Problem Cues, Part 1	1	14
3	Finding Missing Numbers	1	15
4	Finding Patterns in Sequences	3	16
5	Addition/Subtraction Fact Families	6	16
6	Finding Numbers with Odd or Even Digits	10	17
7	"Splitting the Difference"	Inv. 1	18
8	Positive and Negative Numbers	Inv. 1	18
9	Comparing Numbers	Inv. 1	19
10	Abbreviations and Symbols	18	19
11	Reading Clocks	19	20
12	Estimating or Rounding	20	20
13	Measuring Centimeters	Inv. 2	21
14	Naming Fractions/Identifying Fractional Parts	22	21
15	Geometry Vocabulary	23	22
16	Drawing Fractional Parts	26	23
17	Elapsed Time	27	23
18	Multiplication/Division Fact Families	28	24
19	Area and Perimeter Vocabulary	Inv. 3	25
20	Square Roots	Inv. 3	25
21	Writing Numbers	33	26
22	Place Value (Digit Lines)	33	27
23	Reading Inch Rulers	39	28
24	Measuring Liquids and Capacities of Containers	40	28
25	Regrouping Across Zeros	41	29
26	Ways to Show Division	47	29
27	Multiplication (Carrying on Fingers)	48	30

Teaching Strategies for
Saxon Math 5/4 Intervention (continued)

HINT	CONCEPT	INTRODUCE IN LESSON	TEACHING GUIDE PAGE
28	Rate, Part 1	49	31
29	Fraction-Decimal-Percent Manipulatives	Inv. 5	32
30	Percent	Inv. 5	33
31	Word Problem Cues, Part 2	52	34
32	Multiples	55	34
33	Factors of Whole Numbers	55	35
34	Comparing Fractions	56	36
35	Rate, Part 2	60	37
36	Short Division	64	38
37	Reading Metric Rulers	69	39
38	Fraction of a Group, Part 1	70	39
39	Gram/Kilogram Manipulatives	77	40
40	Graphing Coordinates	Inv. 8	41
41	Coordinate Geometry	Inv. 8	42
42	Multiplying by 10, 100, or 1000	85	42
43	Multiplication by Two Digits	87	43
44	Improper Fractions	89	44
45	Roman Numerals	Top. B	45
46	Classifying Quadrilaterals	92	46
47	Fraction of a Group, Part 2	95	46
48	Average	96	48
49	Geometric Solids (Manipulatives)	98	48
50	Faces on a Cube	100	49
51	Probability and Chance	Inv. 10	49
52	Long Division, Part 1	105	50
53	Volume	Inv. 11	51
54	Reducing Fractions (Manipulatives and Shortcuts	112	52
55	Long Division, Part 2	118	52

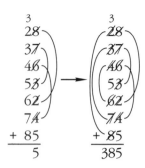

Hint #1:
Column Addition
(Sets of Ten)
(Introduce in Lesson 1)

Students who lack facility with number facts have trouble with column addition. A shortcut is to cross off the sets of ten in the ones column, keeping track of the sets with the fingers of one hand. When all the sets of ten have been found, add the remaining digits and combine this number with the sets of ten. Carry, and then repeat the process with the tens column, the hundreds column, etc.

A "Sets of Ten" chart helps students remember the possible combinations of ten.

For the beginning lessons, it helps to provide a chart of the possible combinations of ten on the board, on the wall, or in an empty area of the Student Reference Guide. *Remind students periodically to cross off sets of ten.* Without a reminder, many will forget to cross off sets of ten and simply continue their slow, self-defeating habits.

Sets of Ten

9 + 1	= 10
8 + 2	= 10
7 + 3	= 10
6 + 4	= 10
5 + 5	= 10

Hint #2:
Word Problem Cues,
Part 1
(Introduce in Lesson 1)

Many students are confused by word problems. Often these students cannot read well and thus do not really understand what is being asked in the question. The textbook identifies word problems as having addition, subtraction, and multiplication patterns.

Teach students to look for keywords in word problems.

As an additional help, students should be taught to look for keywords in word problems. Even if they cannot read the rest of a problem, they might see the word *profit* and know to subtract or see the word *average* and know to divide.

WORD PROBLEM KEYWORDS			
ADD	**SUBTRACT**	**MULTIPLY**	**DIVIDE**
$+$	$-$	$\times$	$\div$
sum total, together, joined (after)	**difference** profit, before, minus; comparisons such as: more than, less than	**product** times, of, cover, double	**quotient** each, per, average

The chart above is located on page 3 in the Student Reference Guide.

Hint #3:
Finding
Missing Numbers
(Introduce in Lesson 1)

Finding missing numbers requires a somewhat sophisticated type of thinking. For some students, the procedure may be visualized by giving examples such as the following:

*"You have some cookies in your pocket. I take three away, and you have two left. How many cookies did you start with? First decide if the amount you started with was **more** or **less** than three. If it was **more**, we **add**. If it was **less**, we **subtract**."*

Students will associate the words *more* with add and *less* with subtract, enabling them to solve the problem.

"Was the amount you started with more or less than three?" (more)

"Then we add. 3 + 2 = 5. You started with five cookies."

Other students cannot perform the missing number calculations unless they use the "Missing Numbers" chart on page 3 in the Student Reference Guide (and below). Usually, with enough practice using this chart, students will internalize the concepts and no longer need to look them up. The most important part of this chart is the two-part subtraction rule.

The most important part of the "Missing Numbers" chart is the two-part subtraction rule.

MISSING NUMBERS	
OPERATION	EXAMPLES
ADDITION: To find the missing **addend** $\longrightarrow$ subtract	$\begin{array}{r} 2 \\ +A \\ \hline 5 \end{array}$ $\begin{array}{r} 5 \\ -2 \\ \hline A=3 \end{array}$ $\begin{array}{r} B \\ +3 \\ \hline 5 \end{array}$ $\begin{array}{r} 5 \\ -3 \\ \hline B=2 \end{array}$
SUBTRACTION: 1. To find the missing top number (minuend) $\longrightarrow$ **add**	$\begin{array}{r} N \\ -3 \\ \hline 2 \end{array}$ $\begin{array}{r} 3 \\ +2 \\ \hline N=5 \end{array}$
2. To find the missing bottom number (subtrahend) $\longrightarrow$ subtract	$\begin{array}{r} 5 \\ -Y \\ \hline 2 \end{array}$ $\begin{array}{r} 5 \\ -2 \\ \hline Y=3 \end{array}$
MULTIPLICATION: To find the missing **factor** $\longrightarrow$ divide	$\begin{array}{r} 3 \\ \times N \\ \hline 6 \end{array}$ $\begin{array}{l} N=2 \\ 3\overline{)6} \end{array}$ $\begin{array}{r} N \\ \times 2 \\ \hline 6 \end{array}$ $\begin{array}{l} N=3 \\ 2\overline{)6} \end{array}$
DIVISION: 1. To find the missing **dividend** $\longrightarrow$ multiply	$2\overline{)N}^{\,8}$ $\begin{array}{r} 8 \\ \times 2 \\ \hline N=16 \end{array}$
2. To find the missing **divisor** $\longrightarrow$ divide	$N\overline{)8}^{\,2}$ $2\overline{)8}^{\,N=4}$

Hint #4:
Finding Patterns
in Sequences
(Introduce in Lesson 3)

Teach students how to inspect the numbers to discover the rule. A cue for this procedure is "Find the pattern. Continue it."

Finding the pattern in a sequence is an automatic task for most students, but students who have little or no facility with numbers must be shown the strategy. Sequences are fairly simple to find using the multiplication table on page 2 in the Student Reference Guide. Knowing which row or column to use, however, is more difficult.

A simple procedure is to subtract any two adjacent numbers in the sequence. Once this difference is found, locate that number's row or column (it does not matter which) in the table. Then determine whether the numbers in the pattern increase or decrease, and fill in the missing numbers. For example, with the problem below, the following dialogue might be used:

"Find the next three numbers in this sequence."

..., 24, 20, 16, _____, _____, _____, ...

"Look at the pattern and find the difference between the first two numbers." (24 − 20 = 4)

"Now find the multiplication table in your reference guide."

"Look at the 4's row (or column) in the multiplication table."

"Do the numbers in the sequence increase or decrease?" (decrease)

"What is the rule?" (Count down by fours.)

"What are the next three numbers?" (12, 8, 4)

For other types of sequences, teach students how to inspect the numbers to discover the rule. Ask "What is done to each number to make the next number?" A cue for this procedure is "Find the pattern. Continue it."

Hint #5:
Addition/Subtraction
Fact Families
(Introduce in Lesson 6)

Pointing out that only a finite number of facts must be learned makes it easier for students to learn those facts.

Students who have difficulty remembering number facts may be overwhelmed by the number of them to be learned. Two things will make this simpler. First, when a relationship of three numbers is shown, the facts are much easier to remember. Second, when students realize only a finite number of facts must be learned, their confidence that they can learn those facts increases. It might be helpful to make a classroom set of triangular flash cards such as the one shown below. Use the following dialogue to introduce addition/subtraction fact families:

"The numbers 5, 6, and 11 form a fact family. Write two addition and two subtraction facts using these three numbers."

```
        5     6    11    11
 /11\   +     +    −     −
/5  6\  __    __   ___   ___
```

*"If you know one fact family, you know **four** facts."*

5 + 6 = 11	11 − 5 = 6
6 + 5 = 11	11 − 6 = 5

Hint #6: Finding Numbers with Odd or Even Digits

(Introduce in Lesson 10)

An "Odd/Even" chart is on page 5 in the Student Reference Guide.

When students are asked to use higher-level thinking skills, they can succeed if given a strategy. A common problem students will be asked to solve in *Saxon Math 5/4* is "Use the digits 1, 2, and 3 once each to write an even number less than 200."

Use the following dialogue to explain how to solve the problem:

"If we use the digits 1, 2, and 3 once each, how many digits will the number have?" (three)

"Draw three digit lines."

— — —

What must the last digit be in order for the number to be an even number?" (2)

"Write the '2' above the last digit line."

— — <u>2</u>

"Which digit must go in the hundreds place to make a number less than 200?" (1)

"Write the '1' above the first digit line."

<u>1</u> — <u>2</u>

"Now write the remaining digit above the remaining digit line."

<u>1</u> <u>3</u> <u>2</u>

The "Odd/Even" chart on page 5 in the Student Reference Guide (and shown below) will be very helpful to students.

ODD/EVEN
Odd numbers: 1, 3, 5, 7, 9, ... Even numbers: 0, 2, 4, 6, 8, ...

Hint #7: "Splitting the Difference"

(Introduce in Investigation 1)

"Split the difference" to find points on number lines and scales and to read temperature.

"Splitting the difference" is a technique used to find points on number lines and scales and to read temperature on thermometers. Use the cue "split the difference" to remind students of the four-step procedure:

1. Subtract the two numbers on both sides of the point to **find the difference.**
2. **Count the parts** between the two numbers.
3. **Split the difference** by dividing the difference by the number of parts.
4. Now **count** by that number.

For example, to find the point indicated below, use the following procedure:

1. Subtract the two numbers on both sides of the point to find the difference.

$$450 - 250 = 200$$

2. Count the parts between the two numbers selected.

3. Split the difference by dividing the *difference* by the number of *parts.*

$$4\overline{)200} \quad \frac{50}{}$$

4. Now count by that number (the quotient found in step 3). Since the quotient was 50, count tick marks by 50's from 250 to the arrow. The arrow points to 400.

Hint #8: Positive and Negative Numbers

(Introduce in Investigation 1)

A discussion of negative numbers in terms of below-zero temperatures will be effective in areas where winter temperatures drop below zero. Most students will also be able to understand the concept of a bounced check. Use the following dialogue:

> *"If you have $3 in the bank and you write a check for $5, you will have a problem. How much will you have to repay the bank?"* (Avoid discussing service charges for overdrafts until a later date.)

Students will also find it helpful to have a number line showing positive and negative numbers that they can work with.

See "Number Line" on page 5 in the Student Reference Guide.

Hint #9:
Comparing Numbers

(Introduce in Investigation 1)

Teach students that the "big" (or "open") end points to the bigger number and vice versa.

Students may need memory cues to remember the symbols for "less than" (<) and "greater than" (>). Since each symbol has a "big" end and a "little" end, demonstrate that the "big" (or "open") end points to the bigger number and vice versa. Some visual learners may benefit from drawing a graphic such as a "hungry alligator" whose mouth opens up to "eat the bigger number."

LESS THAN/GREATER THAN	
15 < 50	50 > 15
little < big	big > little

The chart above is on page 5 in the Student Reference Guide.

Hint #10:
Abbreviations
and Symbols

(Introduce in Lesson 18)

Remind students that units are part of the answer.

Some common math abbreviations and symbols are used throughout the *Saxon Math 5/4* textbook.

centimeter	cm		mile	mi
degree Celsius	°C		milliliter	mL
degree Fahrenheit	°F		millimeter	mm
foot	ft		minute	min
gallon	gal		ounce	oz
hour	hr		pint	pt
inch	in.		pound	lb
kilogram	kg		quart	qt
kilometer	km		second	s
liter	L		yard	yd
meter	m		year	yr

The answers to some problems use a slightly different form or a combination of abbreviations. Alert students to the different ways these abbreviations may be used. The word *units* in an answer box on the worksheets will remind students that the units are actually part of the answer. Some examples follow:

Inv. 3	area expressed in square units	sq. in., square inches; sq. ft, square feet
Lesson 66	miscellaneous label	units
Inv. 11	volume expressed in cubic units	cu. in., cubic inches; cu. cm, cubic centimeters

Hint #11: Reading Clocks

(Introduce in Lesson 19)

Many students have difficulty reading analog clocks, but vocabulary practice will help.

Many students have learned to read only digital time; sometimes even gifted students cannot read analog clocks. This is often due to a lack of familiarity and practice rather than to a learning disability. For example, many students have no idea what "five minutes to eight" means; they only know this time as "7:55." Therefore, a new vocabulary must be taught and practiced.

One suggestion is to ask parents to buy their children inexpensive analog watches. Another is to modify an existing analog wall clock in the classroom (see below). Use cardboard to make a "halo" around the clock; each half of the halo (i.e., from 12 to 6 and from 6 to 12) should be a different color. (The colors make the clock easier to read from a distance.) Mark five-minute intervals on the halo, and place signs next to the 15 and the 45 labeled "quarter after" and "quarter to," respectively. (*Note:* When asked to identify the time "quarter after four," almost every student will initially respond "4:25" instead of "4:15." This is because money is familiar to students, and they know that a quarter equals 25 cents.)

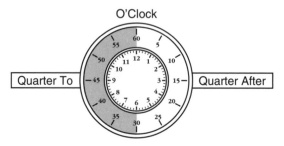

"Learning clocks" are also a good option. Older students may be reluctant to be seen using larger learning clocks, so instead consider purchasing three or four smaller clocks to place around the room. These clocks must be gear-operated so that when the minute hand is moved the hour hand advances appropriately.

Hint #12: Estimating or Rounding

(Introduce in Lesson 20)

This procedure for rounding numbers and decimals will work in any situation.

Estimating amounts is one of the most valuable lifelong skills students can learn. At school, students can use this skill to estimate whether their answers are correct. In other situations, students can estimate how much purchases will cost and whether they have enough money to make those purchases. Most of the time, students will not have number lines available to help them round numbers. The method described below will work in any situation. It shows how to round whole numbers and decimals to any place value.

1. Underline the place value that you will be rounding to (e.g., tens place).

2. Circle the digit to its right. (The circle reminds students that this digit will become a zero.)

3. Ask: Is the circled number 5 or more?

    If so, add 1 to the underlined number.

    If not, the underlined number stays the same.

4. Replace the circled number (and any numbers after it) with zero.

Example: Round 67 to the nearest ten.

$$\underline{6}\,⑦ \quad \longrightarrow \quad 70$$

Hint #13:
Measuring
Centimeters

(Introduce in Investigation 2)

Many students have had little or no experience actually measuring objects; thus, it is important to have metric rulers available in the classroom. Ask students to measure anything nearby—their thumbs, their textbooks, their desks, etc. Students also need a clear idea of the length of a centimeter. Use common objects to give them an idea of how long one centimeter is in comparison to one inch. For example, the width of a large paper clip is about one centimeter, and the diameter of a quarter is about one inch.

Hint #14:
Naming Fractions/
Identifying
Fractional Parts

(Introduce in Lesson 22)

To help instill the concept of fractional parts, recite "___ out of ___ parts" while pointing to a variety of shaded figures.

The main difficulty students have with identifying fractional parts is that they forget the denominator represents the total possible parts. Thus, many students tend to identify the figure below as showing one third (counting only the unshaded parts in the denominator). One solution is to have students recite "___ out of ___ parts" while pointing to the figure. So, to find the fraction represented by the figure below, have students point while saying "This shows **one** out of **four** parts."

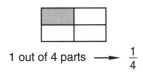

1 out of 4 parts $\longrightarrow$ $\frac{1}{4}$

Repeated practice with the phrase "___ out of ___ parts" should help. Also, remind students that in fractions the **numerator** (top number) is simply the number of shaded parts. The **denominator** (bottom number) is the total number of parts both shaded and unshaded.

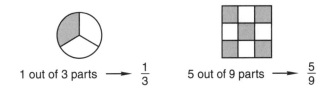

1 out of 3 parts $\longrightarrow$ $\frac{1}{3}$ 5 out of 9 parts $\longrightarrow$ $\frac{5}{9}$

Hint #15:
Geometry Vocabulary

(Introduce in Lesson 23)

Having students use their arms or fingers to demonstrate the concepts of parallel and perpendicular is usually quite effective.

Geometric concepts are not difficult, but the vocabulary sometimes is. Some suggestions for teaching various terminology are described below.

- **Parallel/Perpendicular:** One trick is to point out the parallel *l*'s in the word *parallel* to distinguish it from *perpendicular*. Another is to use an example of railroad tracks to describe the concepts. The best method, however, is to have students use their arms or fingers to demonstrate parallel and perpendicular. Be careful that parallel is not always vertical; this may result in confusion when horizontal and oblique parallel lines are introduced.

- **Intersect:** Use the example of a street intersection. Have students show you how their arms can intersect or be parallel. The students should name the words as they demonstrate them.

- **Right angle:** Tell students that the corner of a rectangular piece of paper will fit "just right" into a right angle. Given a variety of obtuse, acute, and right angles, have students hold each one up to the corner of a piece of paper and tell you whether the angle is too small, too big, or "just right." Gradually, the term "right angle" will assume more meaning for the students.

See "Types of Angles" and "Types of Lines" on page 7 in the Student Reference Guide (also reproduced below).

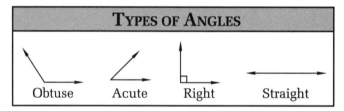

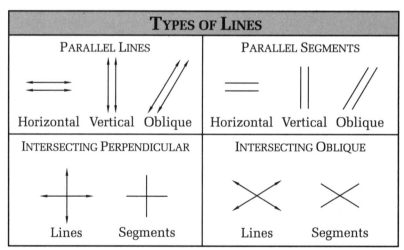

Hint #16:
Drawing
Fractional Parts

(Introduce in Lesson 26)

*Remember the phrase "Fractional parts are **equal** parts."*

Some students find it difficult to understand that fractional parts are **equal parts.** For example, when asked to show $\frac{1}{3}$ of a circle, many students will draw a circle with three (unequal) "stripes" and one stripe filled in.

Students simply need to be shown a strategy to make *equal* parts.

With circles, the key is to begin by drawing a dot in the center of the figure. Thus, to divide a circle in half, put a dot in the center of the circle and draw a dividing line through that dot. To divide a circle into thirds, put a dot in the center and draw a "Y" from the dot. To divide a circle into fourths, divide it in half and then in half again.

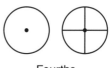

Halves   Thirds   Fourths

Later, students can be shown how to divide figures into even smaller parts. For example, to divide a rectangle into eighths, first divide the figure into two equal parts by drawing a horizontal line through the center. Then divide the rectangle into eight equal parts using vertical lines.

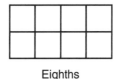

Eighths

Hint #17:
Elapsed Time

(Introduce in Lesson 27)

Since many students are not comfortable with analog clocks, elapsed time is difficult for them. First, students might use "learning clocks" to practice the concept of going forward (clockwise) and backward (counterclockwise). Next, they must become familiar with the vocabulary that tells them when to move clock hands forward, such as *will be,* or when to move clock hands backward, such as *ago.*

Other shortcuts to use when explaining elapsed time are:

- A half hour, or 30 minutes, will be a line directly across the circle (clock) from wherever the minute hand is pointing.
- Visualize the clock divided into four equal parts. Each part is one fourth of an hour, or 15 minutes. (It is useful to name the points "15," "30," "45," and "o'clock.")
- The time twelve hours later is written the same as the present time, except the "a.m." or "p.m." will change.
- The time twenty-four hours later is written exactly the same as the present time, except the day of the week will change.

Hint #18:
Multiplication/
Division Fact Families

(Introduce in Lesson 28)

The poster set, which includes the "Multiplication/ Division Fact Families" poster, is available as an optional purchase.

Remember, two things in particular will help when teaching fact families. First, when a relationship of three numbers is shown, the facts are much easier to remember. Second, when students realize only a finite number of facts must be learned, their confidence that they can learn those facts increases.

It might be helpful to make a classroom set of triangular flash cards such as the ones shown below. Use the following dialogue to introduce multiplication/division fact families:

"Any number times zero is zero."

"Any number times one equals itself."

"Everybody seems to know their 2's and 5's facts."

"These are the only facts left to learn."

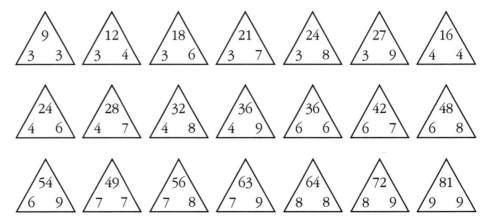

*"If you know one fact family, you know **four** facts."*

3 × 9 = 27	27 ÷ 3 = 9
9 × 3 = 27	27 ÷ 9 = 3

Hint #19:
Area and Perimeter Vocabulary

(Introduce in Investigation 3)

The manipulative kit contains color tiles and cubes to help foster a tactile understanding of area, perimeter, and volume.

The problem of distinguishing between the terms *area* and *perimeter* is often one not of concept but of receptive language. One way to help is to give each student a 4" × 6" index card that will represent a backyard. Each student should draw a "fence" around the yard, using the word *perimeter* repeatedly (see example below). With a green crayon, fill in the "grass" with the word *area*. By using both tactile and visual modalities, students will have a better chance to internalize the concepts and vocabulary. (For students who have difficulty writing, print out multiple copies of the words and have the students cut and paste the words onto the cards.) Students should keep their cards with their reference guides.

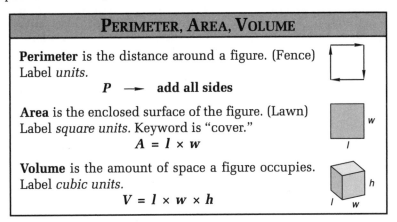

See also "Perimeter, Area, Volume" on page 7 in the Student Reference Guide (also reproduced below).

PERIMETER, AREA, VOLUME

Perimeter is the distance around a figure. (Fence) Label *units*.

$$P \longrightarrow \text{add all sides}$$

Area is the enclosed surface of the figure. (Lawn) Label *square units*. Keyword is "cover."

$$A = l \times w$$

Volume is the amount of space a figure occupies. Label *cubic units*.

$$V = l \times w \times h$$

Hint #20:
Square Roots

(Introduce in Investigation 3)

Use the multiplication table in the Student Reference Guide to help students learn square roots.

In Investigation 3 students are taught how to find the length of a side of a square when the area is known by finding the square root of the area. A shortcut that can be taught uses the multiplication table on page 2 in the Student Reference Guide. Scan the shaded, circled numbers to find the given perfect square (for example, 100). Then look to the top of the circled number's column to find the square root (in this case, 10). This will help students find square roots easily and quickly.

Hint #21:
Writing Numbers
(Introduce in Lesson 33)

Writing numbers correctly will be an important skill for students in later life. One example of using this skill is writing checks. Teach students the following guidelines for writing checks. Instructional Transparency 2 provides blank checks for practice.

- Use hyphens in all numbers between 21 and 99 (except those numbers that end with zero).

- Use a comma after the word *thousand* (or any word greater than one thousand that names place value).

- Use the word *and* only at the end of a whole number to indicate that a fraction or decimal will follow. For example, for $201.13, write "two hundred one and $\frac{13}{100}$ dollars."

- Use the *ths* ending when writing decimal numbers.

A "Spelling Numbers" chart is on page 5 in the Student Reference Guide.

Students may refer to the "Spelling Numbers" chart on page 5 in the Student Reference Guide (and shown below) to help spell numbers correctly.

SPELLING NUMBERS	
WHOLE NUMBERS	FRACTIONS
11 eleven	$\frac{1}{2}$ one half
12 twelve	
13 thirteen	$\frac{2}{3}$ two thi<u>rds</u>
14 fourteen	
15 fifteen	$\frac{3}{5}$ three fif<u>ths</u>
21 twenty-one	
32 thirty-two	$\frac{94}{100}$ ninety-four hundred<u>ths</u>
43 forty-three	
54 fifty-four	$\frac{49}{1000}$ forty-nine thousand<u>ths</u>
65 sixty-five	
76 seventy-six	DECIMALS
87 eighty-seven	
98 ninety-eight	0.1 one ten<u>th</u>
123 one hundred twenty-three	0.94 ninety-four hundred<u>ths</u>
1234 one thousand, two hundred thirty-four	0.049 forty-nine thousand<u>ths</u>

Students often find the most difficult number words to spell are *forty, ninety, nine,* and *four.*

Hint #22:
Place Value
(Digit Lines)

(Introduce in Lesson 33)

Teach students to
discover the pattern in
reading numbers.

Once students discover the pattern of place value, it makes reading and writing numbers much simpler. Point out the "Place Value" chart on page 5 in the Student Reference Guide (reproduced below). Call attention to the obvious groups of three digits separated by commas. If we call these groups families, each family has a different last name: the units (or ones) family, the thousands family, the millions family, etc. Every family has only three members: ones, tens, and hundreds. Commas separate the three-digit families. Starting with the thousands family, we always give the family name to the comma following the family. Point out that the units family does not have a comma. Drawing a circle around each "family" helps to visualize this idea.

(347) , (628) , (407)

Ask students to read only the circled 347. Now ask them to identify the family by the comma. Next have them read the circled 628. Identify the family by the comma. Finally have them read the circled 407. Since there is no comma after it, we stop there. So this number is read as "three hundred forty-seven **million,** six hundred twenty-eight **thousand,** four hundred seven."

PLACE VALUE		
Whole Numbers		Decimals

hundred millions	ten millions	millions		hundred thousands	ten thousands	thousands		hundreds	tens	ones		tenths	hundredths	thousandths
Millions			Thousands			Units (Ones)			$\frac{1}{10}$	$\frac{1}{100}$	$\frac{1}{1000}$			
10^8	10^7	10^6	10^5	10^4	10^3	10^2	10^1	10^0	10^{-1}	10^{-2}	10^{-3}			

Thinking of numbers in groups of three digits will also help students write numbers greater than 999. Using digit lines further simplifies this task. Use the following dialogue to explain the technique:

This approach will
help students write
multidigit numbers.

"Use digits to write 'eighty-two thousand, five hundred three.' "

"How many digits does 'eighty-two' have?" (two)

"Draw two digit lines."

— —

"Now the word thousand *tells you to draw a comma."*

— —,

"Guaranteed, every comma is followed by three digits. Draw three digit lines."

— —, — — —

"Now fill in the digits."

8 2 , 5 0 3

The manipulative kit contains rulers that can be used with the overhead.

Reading inch rulers is an important life skill for students to learn. Make a set of transparent rulers using clear plastic sheets. Each ruler should show different fractional divisions. The rulers may be placed on top of each other to demonstrate equivalencies. (One half is really two fourths, etc.)

Show students how to count fourths on a quarter-inch ruler (e.g., $\frac{1}{4}$, $\frac{1}{2}$, $\frac{3}{4}$, 1, $1\frac{1}{4}$). This will take much practice before it becomes routine. When students are asked to measure items to the nearest quarter inch, it is imperative that they be taught to place the zero mark of the ruler even with the end of the item they are measuring. (This will *not* be obvious to many students!)

There is really no substitute for the use of manipulatives to understand liquid measurements. A set of inexpensive plastic containers (gallon, quart, pint, and cup) can easily be obtained at a discount store; liter bottles are also easy to find. Because working with liquids can be messy, allow students to use pinto beans or other small objects to discover the relationships among various measurements, such as how many cups make a pint.

Once students have a concrete level of understanding of liquid measurements, they will need a "survival tool" to work problems. The chart "Liquids," found on page 1 in the Student Reference Guide (also shown below), will be very useful to students as a tool to work problems.

Have students copy the chart onto notebook paper two or three times. First, write a large G to represent one gallon. Next, write four Q's inside the G to represent the fact that there are four quarts in a gallon. Next, write two P's inside each Q to represent two pints in a quart. Finally, write two c's inside each P to represent two cups in a pint.

Almost every question about nonmetric liquid measurement can be answered using this chart.

LIQUIDS

1 c = 8 oz
1 pt = 16 oz
1 qt = 32 oz

Almost every question about nonmetric liquid measurement can be answered using this chart. The "Equivalence Table for Units" (shown on page 1 in the Student Reference Guide) is also useful, but not nearly as memorable as the large "G."

Hint #25: Regrouping Across Zeros

(Introduce in Lesson 41)

Remember the phrase "Borrow across **all** zeros."

Students tend to get lost when they regroup one step at a time (as many were taught to do in earlier grades). A simpler method is to borrow across all the consecutive zeros and then subtract. Students can usually grasp this process quickly and are far more accurate in their computations. (In money problems, ignore the decimal point when borrowing.)

$$
\begin{array}{r} {\scriptstyle 2\,9\,9} \\[-2pt] {\scriptstyle 1} \\[-4pt] 3001 \\ -\,1322 \\ \hline 1679 \end{array}
\qquad
\begin{array}{r} {\scriptstyle 3\,9} \\[-2pt] {\scriptstyle 1} \\[-4pt] 406 \\ -\,159 \\ \hline 247 \end{array}
\qquad
\begin{array}{r} {\scriptstyle 4\ 9} \\[-2pt] {\scriptstyle 1} \\[-4pt] \$5.00 \\ -\,\$2.34 \\ \hline \$2.66 \end{array}
$$

For the example on the left above, say the following:

"I can't take two from one, so I'll borrow one from 300. This leaves 299. The one I borrowed goes here and becomes eleven. Now let's subtract as usual."

Hint #26: Ways to Show Division

(Introduce in Lesson 47)

The ability to read division problems correctly is critical to students' future ability to simplify improper fractions.

Students often have trouble with the various ways of showing division, largely because they do not understand the concept thoroughly. Many students simply do not hear any difference between "twenty-four divided by six" and "six divided by twenty-four." This is especially true of students who are not developmentally ready to grasp the abstract nature of division. Therefore, although it is certainly not mathematically pure, for the time being tell students to say the greater number first. This idea, coupled with repeated oral and written practice of the various ways to write "twenty-four divided by six," is often the best way to instill the concept of division. For example:

*"Say the **greater** number (dividend) **FIRST.**"*

"twenty-four divided by six" $6\overline{)24}$ $24 \div 6$ $\dfrac{24}{6}$

The ability to read division problems correctly is critical to students' future ability to simplify improper fractions. Students with special needs are sometimes expected to do division before they are developmentally ready, and the procedure described above helps develop a level of awareness sufficient to "survive" until that ability evolves.

The chart below is on page 2 in the Student Reference Guide.

DIVISION
Three ways to show division:
$divisor\overline{)\textbf{dividend}}\qquad \dfrac{\textbf{dividend}}{divisor} = \text{quotient}$
$\textbf{dividend} \div divisor = \text{quotient}$
Example: "Twelve divided by four equals three."
$4\overline{)12}^{\,3}\qquad \dfrac{12}{4} = 3\qquad 12 \div 4 = 3$

Hint #27: Multiplication (Carrying on Fingers)
(Introduce in Lesson 48)

Teaching students to "carry on their fingers" will help later with more complicated math procedures such as long division.

Two-digit multiplication can be confusing to students who have difficulty with task sequencing. The problem is compounded if students have visual perception problems; they often have difficulty lining up figures and cannot achieve accurate computations. Part of the problem is in carrying digits. In problems where several digits are multiplied by two digits, the "carry" numbers tend to get stacked on top of each other. It takes a very organized person to keep the numbers straight.

Therefore, in the early stages of multiplying two digits by one digit, it is best to insist that students learn to carry on their fingers. The process takes a little practice, and you must make sure students are doing this by actually having them show you that they are carrying on their fingers. Another reason for requiring students to carry on their fingers is that in more advanced long division, students must be able to multiply the digits on the quotient line by the divisor. If students are in the habit of carrying on their fingers, this difficult process will be simpler.

To teach the process of carrying on fingers, use the following example:

$$\begin{array}{r} 43 \\ \times\ 7 \\ \hline 301 \end{array}$$

"Multiply the ones digit." $(3 \times 7 = 21)$

*"Write down the **last** digit of that answer."* (Write down the "1.")

*"Carry the **first** digit of that answer on your fingers."* (Hold up two fingers.)

"Multiply the tens digit." $(4 \times 7 = 28)$

"Add the 'carried' digit to that answer." $(28 + 2 = 30)$

Hint #28:
Rate, Part 1

(Introduce in Lesson 49)

By working with this loop method, students will be better prepared to learn about proportion, ratio, and rate.

Proportion and ratio are not formally taught in *Saxon Math 5/4*. However, a simple method can be introduced to help students solve these problems until they are ready for the more algebraic approach. They can also use the procedure to work with unit pricing and conversion of measurement equivalents. *Although this loop method is not in the* Saxon Math 5/4 *textbook, it is very helpful to students with special needs.*

The method is fairly simple. Students must visualize the problem components and know what to do with them. The procedure is as follows:

1. After reading the word problem, identify the two things the problem is about (miles and hours, money and time, inches and yards, etc.). Write these in a column. (It does not matter which item is placed on top.)

2. Carefully fill in what you know. It is important that the known quantities be written directly across from the items they refer to.

3. Carefully write the information you are looking for, placing a question mark or *x* in the "unknown" spot. Again, make sure that the quantities and question mark are written directly across from the appropriate item names.

4. Draw a loop around the numbers that are diagonally opposite. **The loop should never include the question mark.**

5. Multiply the numbers inside the loop. This "answers" the question mark.

(A final step will be introduced in Lesson 60; the process described above works only if the number outside the loop is 1.)

Example:

"If you drive your car at 30 miles per hour, how far will you travel in 4 hours?"

"Name the two things the problem is about."

    mi
    hr

"Fill in what you know."

$$\frac{\text{mi}}{\text{hr}} \quad \frac{30}{1}$$

"Fill in what you're looking for."

$$\frac{\text{mi}}{\text{hr}} \quad \frac{30}{1} = \frac{?}{4}$$

"Make a diagonal loop and multiply."

$$4 \times 30 = 120 \text{ mi}$$

"You'll travel 120 miles in 4 hours."

Hint #29:
Fraction-Decimal-
Percent Manipulatives

(Introduce in Investigation 5)

Fraction, Decimal, and Percent Tower™ cubes can be purchased from ETA/Cuisenaire at (800) 445-5985.

Manipulatives help students grasp the essential relationship among fractions, decimals, and percents. Fraction, Decimal, and Percent Tower™ cubes can be purchased from ETA/Cuisenaire. You can also make your own color-coded and proportionally sized manipulatives to help establish the relationships among fractions, decimals, and percents in the tactile modality.

To make fraction-decimal-percent tower manipulatives, gather scissors, a ruler, and construction paper in a variety of colors.

Cut nine strips of construction paper approximately 12 cm by 2 cm each. Each strip should be the same size but a different color. Using a ruler and scissors, cut eight strips into the following parts: halves, thirds, fourths, fifths, sixths, eighths, tenths, and twelfths. Leave one of the strips whole. When finished, the "whole" strip might be red; the two "$\frac{1}{2}$" strips might be pink; the three "$\frac{1}{3}$" strips might be orange; the four "$\frac{1}{4}$" strips might be yellow; the five "$\frac{1}{5}$" strips might be green; the six "$\frac{1}{6}$" strips might be teal; the eight "$\frac{1}{8}$" strips might be navy blue; the ten "$\frac{1}{10}$" strips might be purple; and the twelve "$\frac{1}{12}$" strips might be black. After cutting the strips, make sure the "parts" equal the "whole"! Label each part with its correct fraction.

Next, make matching sets of decimal and percent strips. Use the same colors chosen for the fraction set. Thus, if the "$\frac{1}{2}$" fraction strips were pink, make the "0.5" decimal strips and the "50%" percent strips pink also. For example:

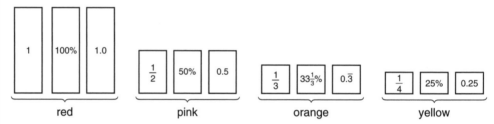

When you are finished, students can use the strips to compare fractions (such as $\frac{1}{4}$ and $\frac{2}{5}$), to compare percents (10% and 20%), and to compare decimals (0.25 and 0.75). They can also use the strips to compare equivalencies ($\frac{1}{4}$, 0.25, and 25%). Whatever the comparison, the color coding serves as a powerful visual memory tool.

Hint #30: Percent

(Introduce in Investigation 5)

The poster set, which includes the "Fraction-Decimal-Percent Equivalents" chart, is available as an optional purchase.

Students may use the tower manipulatives described in Hint #29: Fraction-Decimal-Percent Manipulatives to discover the relationships among fractions, decimals, and percents. Since the towers are color-coded and proportionally sized, students can make a tactile comparison of the two types of towers. For instance, examine the "50%" tower and the "$\frac{1}{2}$" tower. (Both are the same size and color—pink.)

Once students have explored fractions and percents using manipulatives, present the more abstract representation "Fraction-Decimal-Percent Equivalents" found on page 6 in the Student Reference Guide (also shown below).

FRACTION–DECIMAL–PERCENT EQUIVALENTS

Guide students to see that the shaded portion of the towers and the unshaded portion of the towers on the chart add up to 100%. Thus, if 75% of the tower is shaded, 25% is not shaded.

The chart can also be helpful in solving comparison problems. Investigation 5 teaches students to compare fractions and percents to $\frac{1}{2}$ and to 50%. Show students how to compare the height of the given number's tower to the "$\frac{1}{2}$" (or "50%") tower. Use the following dialogue to explain this procedure:

"Compare 48% to $\frac{1}{2}$."

"First, find the fraction $\frac{1}{2}$ on the 'Fraction-Decimal-Percent Equivalents' chart. Notice the height of the tower."

"Would the tower for 48% be taller or shorter than that tower?" (48% is between 40% and 50%, so its tower would be shorter than the "$\frac{1}{2}$" tower.)

"So, which number is greater?" ($\frac{1}{2} > 48\%$)

Hint #31:
Word Problem Cues,
Part 2
(Introduce in Lesson 52)

The concept of "equal groups" is logical to students who can grasp abstract concepts. Those students who need a "survival strategy" will benefit from the keywords listed on page 3 in the Student Reference Guide. One of the crucial keywords is *each*. The word *each* usually means to divide. At this stage of their development, students may not know *why* they divide, and they may not be able to read the rest of the word problem. However, if they can find the word *each* in the problem, they will be able to solve it. Eventually, understanding will come; in the meantime, students can progress in math.

WORD PROBLEM KEYWORDS			
ADD	**SUBTRACT**	**MULTIPLY**	**DIVIDE**
$+$	$-$	$\times$	$\div$
sum total, together, joined (after)	**difference** profit, before, minus; comparisons such as: more than, less than	**product** times, of, cover, double	**quotient** each, per, average

Hint #32:
Multiples
(Introduce in Lesson 55)

Understanding the vocabulary is key to understanding multiples. Students must know the difference between multiples of numbers and factors of numbers. Point out that *multiple* reminds you of multiplication, so use the multiplication table on page 2 in the Student Reference Guide to find multiples.

To help students find the multiples of a number, first instruct them to find the number at the top of the multiplication table. The multiples are the numbers (except zero) in the column under the number. Then demonstrate that the rows of the multiplication table can also be used to find multiples of a number.

"Multiples" on page 2 in the Student Reference Guide (and below) is also a helpful reminder.

MULTIPLES
Multiples of 6: 6, 12, 18, 24, 30, 36, ...
Multiples of 9: 9, 18, 27, 36, 45, 54, ...
LCM of 6 and 9 ⟶ 18
LCM ⟶ Least Common Multiple

Hint #33: Factors of Whole Numbers

(Introduce in lesson 55)

Students should learn to list the factors of whole numbers in numerical order.

It is very important that students learn to list the factors of whole numbers in numerical order. The procedure is as follows:

1. Always *start* with the number 1.

2. Always *end* with the number given.

3. Then find all the other factors of the given number. (Use "Multiplication Table" on page 2 in the Student Reference Guide.)

4. List the factors *in order*. Write each factor only *once*. (Example: The factors of 9 are 1, 3, 9.)

Use the following dialogue to explain the technique:

"Let's list all the factors of 12."

"Write the 'start' and 'end' numbers. Leave space to write numbers between."

1   12

"Look down the 2's column in the multiplication table, and see if you can find a 12. What number times 2 equals 12?" (6)

"Write '2' after the 1. Write '6' **before** *the 12."*

1, 2   6, 12

"Look down the 3's column to find a 12. What numbers should we write?" (3, 4)

1, 2, 3, 4, 6, 12

"The next column we would look down is the 4's. Notice that we have already listed 4 as a factor. This means that we have found all of the factors of 12."

"Look at all the factors. Are they in numerical order?" (yes)

To review the process described above, ask students to find the factors of 24. (1, 2, 3, 4, 6, 8, 12, 24)

Hint #34:
Comparing Fractions

(Introduce in Lesson 56)

The "Basic Fraction Circles" chart is provided in the optional poster set.

The concept of fractions can be difficult for students who have not moved into the realm of abstract thinking. For these students, it is often helpful to use manipulatives.

The "Basic Fraction Circles" chart shown below (and included in the poster set) is helpful in teaching students to compare fractions. Students can make their own pie chart manipulatives using Activity Masters 15–17, which are found in *Saxon Math 5/4 Assessments and Classroom Masters*. These manipulatives will help students compare fraction parts to see which are larger or smaller.

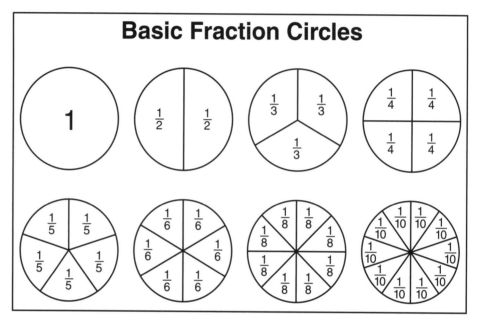

Students may have difficulty distinguishing between the $\frac{1}{5}$ and $\frac{1}{6}$ sections in the pie charts. Using fraction tower manipulatives such as those described in Hint #29: Fraction-Decimal-Percent Manipulatives may make the differences in fractions easier to see.

Students will find the following strategy for comparing fractions particularly useful, especially when fraction manipulatives are not likely to be found.

Example: Compare $\frac{1}{2} \bigcirc \frac{1}{3}$

Cross multiply:

Then compare the numbers on top: $3 > 2$, so $\frac{1}{2} \bigcirc\!\!> \frac{1}{3}$

Hint #35: Rate, Part 2

(Introduce in Lesson 60)

As explained in Hint #28: Rate, Part 1, the simple loop method helps students with proportions, ratios, rate, unit pricing, and conversion of measurement equivalents. The final step in the process occurs when the number outside the loop is not 1. When the number is not 1, divide the "loop" answer by the number outside the loop.

When teaching the final step, use the following approach:

"In Lesson 49, we learned to: Name the two things the problem is about. Fill in what we know. Fill in what we're looking for. Make a diagonal loop and multiply."

"Now we add one last step: **Divide** *the loop answer by the number* **outside** *the loop."*

Example:

"If Joe reads 2 pages per minute, how long will it take him to read 18 pages?"

pages
minutes

$$\frac{2}{1} = \frac{18}{?}$$

$$2\overline{)18}^{\ 9 \text{ minutes}}$$

"It will take Joe 9 minutes to read 18 pages."

See "Proportion (Rate) Problems" on page 4 in the Student Reference Guide (also shown below).

PROPORTION (RATE) PROBLEMS		
EXAMPLE 1	**EXAMPLE 2**	
If the number outside the loop is **1**: $\frac{30}{1} = \frac{x}{4}$	If the number outside the loop is **not 1**: $\frac{3}{5} = \frac{6}{w}$	
Cross multiply. $30 \cdot 4 = x$ $x = 120$	• Cross multiply. $3w = 30$ • Divide by known factor. $30 \div 3 = 10$	

Hint #36:
Short Division

(Introduce in Lesson 64)

Teaching students short division will help improve their mental math ability.

Students frequently get lost in the steps of long division. Although they will need to work all the steps with multidigit divisors, they will be more confident if they have learned to use short division with single-digit divisors.

The long division method is introduced in Lesson 105. Rather than teach the long method, take time to teach the short method described below. Students who have struggled with long division may resist the idea at first, but most will eventually like the new method because it is so much easier. An additional benefit is an improvement in students' ability to do mental math.

The rules of short division are simple:

- Any number "left over" goes in front of the next digit.
- Any final number "left over" becomes the remainder. (Remainders must be smaller than divisors.)
- There must be one digit in the quotient above each digit in the dividend. A cue for this rule is "digit above each digit."
- If necessary, have students use zero as a placeholder.[†] (If the divisor won't divide into the first digit of the dividend, place a small zero above that digit.)

For example, when giving the problem below, simply remind students that after multiplying and subtracting, the "leftover" goes in front of the next digit.

$$3\overline{)8^27}\quad\begin{smallmatrix}2\ 9\end{smallmatrix}$$

Remind students to place a digit above each digit and to use zero as a placeholder. You can also help students visualize the process by circling the first two digits of the dividend (when appropriate).

$$3\overline{)\textcircled{23}{}^24}\quad{}_07\ 8 \qquad 3\overline{)\textcircled{23}{}^25}\quad{}_07\ 8\,\text{R}\,1$$

Insist that students use short division—do not allow the use of long division for single-digit divisors. One reason to insist on short division is that students often have a difficult time lining up numbers correctly in long division. This becomes a stumbling block to the concept of division. Students will cope better with long division if they have confidence in short division.

[†] It might be necessary to remind students to remove initial placeholder zeros when writing their final answers.

Hint #37:
Reading Metric Rulers

(Introduce in Lesson 69)

Metric rulers that can be used with the overhead are available in the manipulative kit.

Hint #38:
Fraction of a Group, Part 1

(Introduce in Lesson 70)

A cue for this strategy is "Divide by the denominator."

Point to one centimeter on a metric ruler. Explain that there are ten tiny millimeters in that centimeter. Ask the students how many millimeters are in two centimeters, then five centimeters, then nine centimeters, etc. Ask the students what factor they are multiplying by. (10)

Now point to the seven-centimeter mark and ask how many centimeters are in seventy millimeters. Continue asking questions such as "How many centimeters are in forty millimeters?" or "How many centimeters are in twenty millimeters?" Ask the students what factor they are dividing by. (10) Tell them they have just learned to speak metric!

In problems about a fraction of a group, the whole number is divided by the number of parts indicated by the denominator. For example, $\frac{1}{4}$ of 12 really means to divide 12 into four parts and consider one of those parts. This can be easily demonstrated using plastic tokens or chips; the demonstration will be necessary for students who are at the concrete level of understanding fractions. Students who are at a more abstract level of understanding can use the method described in the textbook.

The following "survival strategy" can be used both by students who cannot grasp the concept of fractions and by students who are ready to move beyond drawing pictures of fractions:

- $\frac{1}{2}$ of a number $\longrightarrow$ divide by 2

- $\frac{1}{3}$ of a number $\longrightarrow$ divide by 3

- $\frac{1}{4}$ of a number $\longrightarrow$ divide by 4

A cue for this strategy is "Divide by the denominator."

Hint #39: Gram/Kilogram Manipulatives

(Introduce in Lesson 77)

Use common classroom items to compare one gram to one kilogram.

Teaching metric weights is much easier with manipulatives. Although students in the United States usually have some sense of how much a pound of apples weighs, they usually have no such sense of how much a kilogram of apples weighs. Manipulatives will help.

One suggested activity is to use common classroom items to compare a one-gram mass weight (such as a paper clip) to a one-kilogram mass weight (such as a 2.2-pound book). Explain that the kilogram weighs 1000 times the gram. Ask students if they can think of any other items that weigh about a kilogram. Gram and kilogram weights can be purchased from ETA/Cuisenaire.

Also point out the list of metric conversions in the "Equivalence Table for Units" on page 1 in the Student Reference Guide (also shown below). Although the table helps, the activity is more likely to remain in students' memories.

EQUIVALENCE TABLE FOR UNITS	
LENGTH	
U.S. Customary	Metric
12 in. = 1 ft	10 mm = 1 cm
3 ft = 1 yd	1000 mm = 1 m
5280 ft = 1 mi	100 cm = 1 m
1760 yd = 1 mi	1000 m = 1 km
WEIGHT	**MASS**
U.S. Customary	Metric
16 oz = 1 lb	1000 g = 1 kg
2000 lb = 1 ton	1000 mg = 1 g
CAPACITY (LIQUID MEASURE)	
U.S. Customary	Metric
16 oz = 1 pt	1000 mL = 1 L
2 pt = 1 qt	
4 qt = 1 gal	

There are **no common fractions** in the **metric system.** Use **decimals.**

Hint #40:
Graphing Coordinates
(Introduce in Investigation 8)

Graphing coordinates is not a difficult concept, but it is easier to learn with a little practice.

1. Show students that they should first move horizontally and then move vertically to graph points on a coordinate grid. On an overhead transparency or the board, draw and label five horizontal units and five vertical units (with origin at "0" in the lower left-hand corner) as shown below.

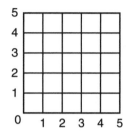

   Use the following dialogue:

   "Let's graph the point (2, 3). What does the '2' tell me to do?"

   Show students that they should first move horizontally two units to the right of zero.

   "What does the '3' tell me to do?"

   Demonstrate by moving vertically three units, and show students how to graph the point. (Make sure students see that you moved vertically from the position of the horizontal number, not the origin.) Repeat (using different points) until students understand that they should always move horizontally first and vertically second when graphing coordinates.

2. If time allows, have the students play hide-and-seek using coordinates. Pair the students and provide each with a five-by-five coordinate grid drawn on paper. Have everyone secretly mark a point on his or her grid as a hiding place. Pairs then take turns trying to find each other's hiding place by guessing four different sets of coordinates. Both students mark the guesses on the coordinate grid. The first one to hit the hiding place of the other wins the game.

Hint #41:
Coordinate Geometry
(Introduce in Investigation 8)

On a piece of graph paper, numerically label five horizontal units and five vertical units with the origin at the lower left-hand corner. Make multiple copies of this coordinate grid worksheet and store them in a convenient, accessible location. When students want to play the hide-and-seek game described in Hint #40: Graphing Coordinates, instruct them to use one of these worksheets.

Hint #42:
Multiplying by
10, 100, or 1000
(Introduce in Lesson 85)

When multiplying a whole number by 10, 100, or 1000, simply add the same number of zeros to the number being multiplied.

Students usually appreciate shortcuts. The following procedure is not difficult and will save students much computation time.

When multiplying a whole number by 10, 100, or 1000, simply add the same number of zeros to the number being multiplied. For example, if multiplying 37 by 100, add two zeros to 37 (for the two zeros in 100). With practice, students should be able to do this procedure as mental math rather than on paper.

The procedure may be summarized as:

- If multiplying by **10**, add **1** zero after the whole number being multiplied.
- If multiplying by **100**, add **2** zeros after the whole number being multiplied.
- If multiplying by **1000**, add **3** zeros after the whole number being multiplied.

When multiplying a decimal number by 10, 100, or 1000, simply move the decimal point to the **right** the same number of places as zeros.

Example: Multiply 1.23 by 100.

There are two zeros in the factor 100, so move the decimal point two places to the right.

$$1.23 \times 100 = 123$$

Hint #43: Multiplication by Two Digits

(Introduce in lesson 87)

When multiplying by two digits, visualize the two-digit multiplier as two one-digit multipliers.

Multiplication by two digits is difficult for students who have difficulty with sequencing or visual perception. Although most students will be able to multiply two digits by one digit with proficiency by now, many will get confused when multiplying two digits by two digits. The following procedure was developed by a student who also experienced these difficulties.

Visualize the two-digit multiplier as two one-digit multipliers. When finished multiplying by the "first" digit, cross it out; then place an ✗ on the second line of multiplication as a reminder to indent. (Using zeros as placeholders is less effective, since students may confuse the placeholders with problem numbers.) Remind students to carry on their fingers when appropriate.

For the problem 34 × 12, the procedure can be summarized in four steps:

1. Multiply by the ones digit (ignore the tens digit):

$$\begin{array}{r} 34 \\ \times\ 12 \\ \hline 68 \end{array}$$

2. Cross out the 2 when you are finished with it:

$$\begin{array}{r} 34 \\ \times\ 1\cancel{2} \\ \hline 68 \end{array}$$

3. Indent the next line using ✗ as a placeholder, and multiply by the tens digit:

$$\begin{array}{r} 34 \\ \times\ 1\cancel{2} \\ \hline 68 \\ +\ 34\,✗ \\ \hline \end{array}$$

4. Add the two answers:

$$\begin{array}{r} 34 \\ \times\ 1\cancel{2} \\ \hline 68 \\ +\ 34\,✗ \\ \hline 408 \end{array}$$

Graph paper can be used as a visual aid to line up columns and rows.

On problems involving money, it may be necessary to remind students to add dollar signs and decimal points to their final answers.

Hint #44:
Improper Fractions

(Introduce in Lesson 89)

Use the cue " 'top heavy' fractions" when referring to improper fractions.

Many students with math difficulties do not comprehend the equivalencies of mixed numbers and improper fractions. First, because the vocabulary can be confusing, try including the description "top heavy" when referring to improper fractions. Fraction tower manipulatives will help demonstrate that $\frac{5}{4}$ is the same amount as $1\frac{1}{4}$. Students will need much tactile practice with this.

Next, demonstrate the concept with two-dimensional pictures such as those shown below. When drawing the figures showing the improper fraction, divide each figure into the same number of parts as the denominator.

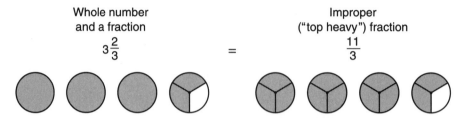

Whole number and a fraction $3\frac{2}{3}$ = Improper ("top heavy") fraction $\frac{11}{3}$

Finally, teach the pencil-and-paper approach. Multiply the denominator by the whole number; then add the numerator. Write this answer over the original denominator.

$$3\frac{2}{3} \longrightarrow 3 \begin{smallmatrix} + \\ 2 \\ \times \\ 3 \end{smallmatrix} = \frac{(3 \times 3) + 2}{3} = \frac{11}{3}$$

Hint #45:
Roman Numerals

(Introduce in Appendix Topic B)

Because Roman numerals can be quite puzzling for students with visual discrimination problems, it is fortunate that Roman numerals are not critical to understanding math. However, with a little manipulation of the spacing between the numbers, you can make them easier to decode. It will be helpful to make the following classroom sets of Roman numeral flash cards:

Set 1: I (1), V (5), X (10), L (50), C (100), D (500), and M (1000)

Set 2: IV (4), VI (6), IX (9), and XI (11)

Set 3: X IX (19), X VI (16), XX IV (24), CD XX V (425), and L IV (54)

Notice the spacing in the third set of flash cards. The cards should include extra space around numerals when two Roman numerals are used to represent one digit in a multidigit number. For example, in the Roman numeral XIX, two numerals (IX) are used to represent one digit (9). To help students decode the Roman numeral XIX, space the numerals to show the value of its parts:

| X | IX |
| 10 | 9 |

The following exercise will help students become comfortable with Roman numerals:

1. First, practice with the first set of flash cards.

2. Once students become familiar with the basic numerals from the first set, practice with the second set of flash cards. Have students try to verbalize the rule they discover about the second set of numbers. If students cannot verbalize the rule, then tell it to them. *If the greater value is written in front of the smaller value,* **add** *the values. If the smaller value is written in front of the greater value,* **subtract** *the values.*

3. Finally, have students practice with the third set of flash cards.

4. Continue to review the three sets of numbers.

With this preparation, students should be able to do the practice for Appendix Topics B and C.

X IX is much easier for students to decode than XIX.

Hint #46: Classifying Quadrilaterals

(Introduce in Lesson 92)

Review the qualities of each quadrilateral by asking "Is every ...," "Are some ..." type questions.

Questions such as "Is every square a rhombus?" require a very high level of abstract thinking. It is imperative to use the "Quadrilaterals" chart on page 8 in the Student Reference Guide (also shown below). Review the qualities of each quadrilateral with students. Point to the chart and ask "Is every ...," "Are some ..." type questions. Also have students ask those types of questions.

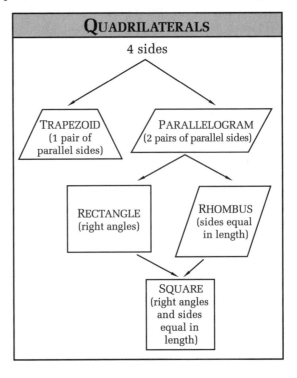

Hint #47: Fraction of a Group, Part 2

(Introduce in Lesson 95)

In Hint #38: Fraction of a Group, Part 1, students learned two ways to solve problems about a fraction of a group:

1. Divide plastic tokens or chips into groups.

2. Divide by the denominator of the fraction.

Often, the numerator will not be 1 in problems about a fraction of a group. Plastic tokens or chips may still be used to provide a tactile demonstration of the concept. For example, show that $\frac{2}{3}$ of 12 is 8 like this:

$$XXXX \quad XXXX \quad XXXX$$

12 is divided into 3 groups.
2 of these 3 groups contain a total of 8.

Another method (more useful with larger numbers) is to have students draw boxes as in the following example:

"Debbie scored two thirds of the 36 points in the game. How many points did she score?"

36 points

$\frac{2}{3}$ Debbie scored. $\left\{ \begin{array}{c} \boxed{12 \text{ points}} \\ \boxed{12 \text{ points}} \end{array} \right.$

$\frac{1}{3}$ Debbie did not score. $\left\{ \boxed{12 \text{ points}} \right.$

"Debbie scored 2 × 12 points, which is 24 points."

Finding a fraction of a number is easily done with the loop method.

Another way to solve this problem is to use the loop method previously learned in Hint #35: Rate, Part 2.

Set up the problem using an "is over of" format:

$$\frac{is}{of} \quad — = —$$

The question can be reworded, "What **is** $\frac{2}{3}$ **of** 36?" Fill in the known numbers, using a question mark for the unknown number. Then draw a loop around the numbers that are diagonally opposite. The loop should never include the question mark.

$$\frac{is}{of} \quad \frac{2}{3} = \frac{?}{36}$$

Finally, multiply the two numbers inside the loop ($2 \times 36 = 72$), and divide by the number outside the loop ($72 \div 3 = 24$). Debbie scored 24 points.

See "Finding a Part When the Whole is Known" on page 4 in the Student Reference Guide (also shown below).

FINDING A PART (FRACTION OR PERCENT) WHEN THE WHOLE IS KNOWN
(Alternate Method)

Set up an "is/of" proportion:

• 2/3 of 600 is what number?

$$\frac{is}{of} \quad \frac{2}{3} = \frac{?}{600} \qquad (600 \cdot 2) \div 3 = \mathbf{400}$$

• 30% of 20 is what number?

$$\frac{30}{100} = \frac{3}{10} \qquad \frac{is}{of} \quad \frac{3}{10} = \frac{?}{20} \qquad (20 \cdot 3) \div 10 = \mathbf{6}$$

Shortcut: Reduce the ratio before multiplying the numbers in the loop.

Hint #48:
Average

(Introduce in Lesson 96)

To compute average, add the numbers and divide by the number of items.

The concept of average is not too difficult to convey. It can be demonstrated effectively with manipulatives. For example, give the students twelve plastic tokens divided unevenly into two (or three) groups. Ask the students to rearrange the tokens in two (or three) equal groups. The number in each equal group is the average.

Another method to use for the textbook problems is more abstract, but still fairly simple. Have the students add the numbers and then divide by the number of items. Tell the students that the answer to any average problem will always be between the smallest number and largest number and that this is a way to "spot check" answers. An example of the procedure follows:

1. To find the average of 24, 26, and 28, first add the numbers:

$$\begin{array}{r} \overset{1}{24} \\ 26 \\ +\ 28 \\ \hline 78 \end{array}$$

2. Next, divide the sum (78) by the number of items (3):

$$3\overline{)7^18}\ \ \overset{26}{}$$

3. Check the answer (26). The answer must be between the smallest number (24) and the largest number (28).

A reminder of the procedure is provided on page 2 in the Student Reference Guide (and below).

AVERAGE
Average → add numbers; then divide.
Halfway → add numbers; then divide by two.

Hint #49:
Geometric Solids
(Manipulatives)

(Introduce in Lesson 98)

The ETA/Cuisenaire Relational GeoSolids set is available in the manipulative kit.

There is really no substitute for manipulatives. Useful ones to have are a cube, a rectangular prism, a square-based pyramid, a triangular-based pyramid, a triangular prism, a sphere, a cone, and a cylinder.

Students will be asked to count the faces, edges, and vertices of these solids. Because many students get lost when counting the number of faces, it helps to paint or put a colored sticker on at least one side of each solid. A better alternative is to buy the ETA/Cuisenaire Relational GeoSolids set, which is available in the manipulative kit. These clear plastic manipulatives are green on one side, helping students keep track of the number of faces. The set also includes a triangular-based pyramid, which is often difficult to find.

The more tactile experiences students have, the sooner they will be able to move to more abstract mathematical thinking.

Hint #50:
Faces on a Cube

(Introduce in Lesson 100)

Enhance students' conception of spatiality by making a metric box.

Although it is possible to simply tell students that cubes have six faces, this would require that they memorize yet another fact; it is more effective to enhance their conception of spatiality. One way to do this is to make a cube called a *metric box*. Start with a sheet of paper or thin cardboard that is 30 cm by 40 cm (about 12 inches by 16 inches). Divide the sheet into twelve 10-by-10-cm squares. Now cut out the "t" shape and fold as shown below.

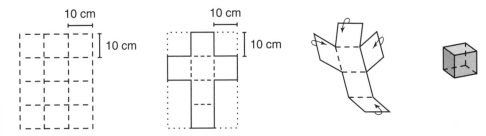

Each edge of the completed cube should be 10 cm long. Each face should be 10 cm by 10 cm. This cube would hold one liter of water. The weight (mass) of the water in this cube would be one kilogram.

If it is not possible for every student to make a cube, consider laminating and creasing a "master" cube for class use. Students will want to fold and unfold the cube many times as they discover that the t-shaped paper forms a cube.

Hint #51:
Probability and
Chance

(Introduce in Investigation 10)

A set of spinners for the overhead is available in the manipulative kit.

Probability and chance express the same concept using different vocabulary. **Probability** uses **fractions**, whereas **chance** uses **percents**. At this point in the book, these should not be difficult concepts. A strategy for explaining them follows:

1. Get a die and a spinner with numbers on it. Ask students to think of some games that use spinners.

2. Spin the spinner and ask students to name the probability that the spinner will stop on a particular number. Explain that *probability* means that there are _____ out of _____ possibilities that the spinner will stop on a particular number.

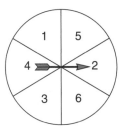

- If it is **certain** to happen, the probability is 1.

  Example: *"What is the probability that this spinner will stop on a number less than 7?"* (1)

- If it is **impossible,** the probability is 0.

  Example: *"What is the probability that this spinner will stop on the number 8?"* (0)

- All other probabilities are stated as fractions (which should be reduced, if possible).

  Example: *"What is the probability that this spinner will stop on an even number?"* $\left(\frac{3}{6}, \text{ which reduces to } \frac{1}{2}\right)$

Repeat the procedure using the die. Continue to repeat the procedure, alternating between chance and probability vocabulary.

Hint #52:
Long Division, Part 1

(Introduce in Lesson 105)

To make long division easier:
- *Use zero as a placeholder.*
- *Place a digit above each digit.*

With two-digit divisors, use long division. (This is the conventional method of "divide, multiply, subtract, and bring down.") Although the sequencing and visual perception demands of long division might be difficult for some students, two things will make the process a little easier and thus more accurate.

First, continue to use zero as a placeholder.[†] Second, remember that there must be a digit in the quotient above each digit in the dividend, including any zeros used as placeholders. Thus, a three-digit dividend must have a three-digit quotient.

The rules may be summarized as follows:

- Use **long division** with **two-digit divisors.** (Continue to use short division with one-digit divisors.)
- "Divide, multiply, subtract, and bring down."
- Use zero as a placeholder.
- Place a digit above each digit.
- Make sure any remainder is smaller than the divisor.

$$
\begin{array}{r}
034\text{R}2 \\
10\overline{)342} \\
-30\downarrow \\
\hline
42 \\
-40 \\
\hline
2
\end{array}
$$

[†]It might be necessary to remind students to remove initial placeholder zeros when writing their final answers.

Hint #53:
Volume

(Introduce in Investigation 11)

Colored cubes, available in the manipulative kit, are an effective way to teach the concept of volume.

Volume is best taught by using manipulatives. Use the colored cubes in the manipulative kit (or sugar cubes). Once students have mastered the concept using objects, it will be easy to use the formula $V = l \times w \times h$. Show students the formula for volume on page 7 in the Student Reference Guide.

Example: Find the number of 1-cm cubes that can fit in the box shown.

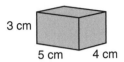

To find the volume of a rectangular prism:

1. Find the number of cubes in one layer of the rectangular prism. (This equals the area of the *base* of the prism.)

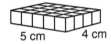

one layer

$5 \times 4 = 20$ cubes

$\dfrac{\text{layers}}{\text{cubes}} \quad \dfrac{1}{20}$

2. Multiply the number of cubes in one layer by the number of layers in the prism (the *height* of the prism).

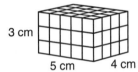

three layers

$\dfrac{\text{layers}}{\text{cubes}} \quad \dfrac{1}{20} = \dfrac{3}{?}$

$20 \times 3 = 60$ cubes

Remind students that volume may be expressed different ways (using the answer in the above example):

60 cu. cm or 60 cubic centimeters or 60 cm^3

Hint #54:
Reducing Fractions
(Manipulatives and
Shortcuts)
(Introduce in Lesson 112)

Working with tactile manipulatives is really the best way for students to internalize the concept of reducing fractions. With manipulatives students can easily see that $\frac{2}{4}$ equals $\frac{1}{2}$, etc. The need to work with manipulatives before attempting pencil-and-paper work cannot be emphasized enough.

The next step is pencil-and-paper work. Simply divide both the numerator and the denominator by the same number. Once the students grasp how to reduce the fractions, it's a good idea to let them refer to the "Fraction Families Equivalent Fractions" chart on page 6 in the Student Reference Guide (and shown below). By doing so, they will be able to commit the equivalent fractions to memory much more quickly. That is, the more frequently students look at the chart, the sooner their visual memories will take over, and the sooner information will come automatically.

By referring to this chart frequently, students will soon be able to commit the facts to memory.

FRACTION FAMILIES
EQUIVALENT FRACTIONS

$\frac{0}{2}$					$\frac{1}{2}$						$\frac{2}{2}$
$\frac{0}{3}$			$\frac{1}{3}$				$\frac{2}{3}$				$\frac{3}{3}$
$\frac{0}{4}$		$\frac{1}{4}$			$\frac{2}{4}$			$\frac{3}{4}$			$\frac{4}{4}$
$\frac{0}{5}$	$\frac{1}{5}$		$\frac{2}{5}$			$\frac{3}{5}$			$\frac{4}{5}$		$\frac{5}{5}$
$\frac{0}{6}$	$\frac{1}{6}$		$\frac{2}{6}$		$\frac{3}{6}$		$\frac{4}{6}$		$\frac{5}{6}$		$\frac{6}{6}$
$\frac{0}{8}$	$\frac{1}{8}$	$\frac{2}{8}$	$\frac{3}{8}$		$\frac{4}{8}$	$\frac{5}{8}$	$\frac{6}{8}$		$\frac{7}{8}$		$\frac{8}{8}$
$\frac{0}{9}$	$\frac{1}{9}$	$\frac{2}{9}$	$\frac{3}{9}$	$\frac{4}{9}$	$\frac{5}{9}$	$\frac{6}{9}$	$\frac{7}{9}$	$\frac{8}{9}$			$\frac{9}{9}$
$\frac{0}{10}$	$\frac{1}{10}$	$\frac{2}{10}$	$\frac{3}{10}$	$\frac{4}{10}$	$\frac{5}{10}$	$\frac{6}{10}$	$\frac{7}{10}$	$\frac{8}{10}$	$\frac{9}{10}$		$\frac{10}{10}$
$\frac{0}{12}$	$\frac{1}{12}$	$\frac{2}{12}$	$\frac{3}{12}$	$\frac{4}{12}$	$\frac{5}{12}$	$\frac{6}{12}$	$\frac{7}{12}$	$\frac{8}{12}$	$\frac{9}{12}$	$\frac{10}{12}$ $\frac{11}{12}$	$\frac{12}{12}$

Hint #55:
Long Division, Part 2
(Introduce in Lesson 118)

Lesson 118 begins giving two-digit divisors that are **not** multiples of ten. The easiest method is to have students round **down** such divisors (as well as the dividends) to the nearest multiple of ten before making the first "guess." For example:

$$\overset{0\,0}{32\overline{)128}} \quad \text{(Think: } 30\overline{)120} \text{)}$$

Estimating the first digit of a quotient is the most difficult level of progression in the entire process of teaching division. Reassure students that understanding will come. If the process continues to be a stumbling block, provide more practice "rounding down" divisors. Students must become comfortable doing this before attempting the task of "rounding up."

Classroom Management

Establish a consistent classroom procedure as soon as possible. Prevention of problems is key!

Early in the school year, establish a consistent classroom procedure that allows students to leave their seats to turn in assignments and to get the materials they need to continue working. Prevention of problems is the key. To help achieve this goal:

- Store worksheets in a location easily accessible to the students (if students do not have individual copies of the Student Workbook).
- Establish a route for students to move around the classroom.
- Store manipulatives in a location easily accessible to the students.
- Use boxes labeled "Work to be corrected" and "Work that's been corrected."
- Instruct students to work on easier problems while waiting for help on more difficult ones.
- If you use self-paced instruction, direct students to begin the next assignment while they wait to receive an assignment that they might need help to correct.
- Designate someone to act as a helper if you are interrupted or called to the door for a conference.
- Establish a three-step plan for handling disruptive student behavior (for example, verbal warning, time out, and detention).

Classroom Aides

Classroom aides create a more effective learning/teaching environment.

In many classrooms, effectively using class time while meeting individual instructional needs is more easily achieved with the assistance of classroom aides (paraprofessionals, teaching assistants, parent volunteers, or students earning elective credits). Classroom aides free you to:

- give one-on-one instruction of new increments
- respond quickly to students who ask for help
- teach appropriate classroom behavior privately
- have regular, quiet, personal interaction with each student
- work on positive, continual reinforcement

The ideal student classroom aides are those who have completed at least one year of Saxon Math™ and have proven themselves to be accurate and reliable. These students can be assigned to help other students.

Adult assistants or parent volunteers can supervise the student classroom aides, keep records, prepare materials for future assignments, help answer students' questions, alert you to potential problems (such as a need for reteaching), and return papers to students[†]. It is recommended that adults perform the more sensitive, managerial tasks and that students be limited to helping with supplies and assisting other students with simple tasks. With efficient classroom aides, you are free to concentrate on meeting individual learning needs.

Motivation

Daily verbal encouragement and positive remarks on graded papers are extremely important to student success. It is amazing how a little shared humor or a reminder of a previous achievement can brighten a student's

---

[†]Local privacy laws might limit the kinds of tasks parent volunteers may perform.

outlook. Take the time to recognize small successes. Reward struggling students for completing and correcting all assignments for the week with a certificate, a pencil, or a small snack. Completing a textbook is a particularly important achievement and should be marked with a small celebration and recognition from the school principal.

Using red ink to check assignments has a negative effect on student morale. Blue and green are better choices. In addition, mark papers with the number of problems worked correctly rather than the number missed.

Involving Parents

Daily assignments are essential for struggling students. Parents should be alerted that homework will be a common occurrence in their child's math education. If homework is not completed, the student will not progress, and his or her grades will suffer. If poor homework habits occur on a regular basis, contact the parents to determine the cause. Call home to share good news as well.

Rules for Success

Students with difficulties will be more likely to succeed in *Saxon Math 5/4* if:

- **they complete the lessons in sequence.** The problem sets have been carefully crafted to provide continual practice and review of all the concepts, processes, and procedures introduced since the first lesson. Students must complete all the problems in each problem set, and the lessons must be completed in sequence.

- **they correct all errors in the lessons.** Because a problem that is missed in one lesson will reappear in another form in subsequent lessons, students must correct their own mistakes to be successful.

- **they complete homework assignments.** Homework is essential. It develops appropriate study habits and makes it possible to close the gap between students' functional level and target grade level.

- **they complete tests in class.** Tests provide you with the information needed to diagnose whether students have mastered the material. If a test indicates that a student has not mastered certain concepts, use the remediation advice on pages 55–58 to determine the next steps.

- **calculator use is limited.** As a general rule, calculators should not be used unless special individual circumstances warrant otherwise. Mathematical skills are maintained through practice in whole-number and decimal calculations. In addition, students demonstrate understanding by showing all the steps required in working problems. Students may use calculators at the teacher's discretion, but students should still work out all problems in the lessons. Calculators should not be allowed during tests.

- **positive reinforcement is used.** Provide continual, positive reinforcement for students' achievements, large and small. With each success, students will gain more confidence and will be more likely to meet future challenges in the mathematics classroom and in other academic areas with positive attitudes and feelings of competence. Students will soon learn to expect success, and their improved self-images will lead to greater effectiveness.

- **different teaching approaches are used.** Because students learn in a variety of ways, use varied approaches to the same concept to reach the greatest number of students. For auditory learners, orally explain concepts as clearly as possible. For visual learners, employ visual aids such as charts, diagrams, and the Student Reference Guide. For kinesthetic learners, have manipulatives readily available.

Remediation

Remediation is integral to ensuring success for all learners. Saxon programs have an intrinsic advantage in the area of remediation: By distributing practice over time, Saxon programs prevent students from forgetting what they have learned. The greater challenge is to help students who do not "get it" after even a week or two of practice.

Diagnosing Gaps in Knowledge

Tests are the best indicators of whether students have misunderstandings. Scores below 80 percent usually signal the need for remediation. The Test-Item Analyses and the Recording Forms in the *Saxon Math 5/4 Intervention* Masters will help in evaluating performance. These materials provide three options for monitoring problems and diagnosing errors: analysis of individual performance, analysis of whole-group performance, and student self-analysis.

Analysis of Individual Performance

The Test-Item Analyses allow you to analyze individual students' performances. They identify the concepts of each test problem and the lessons that introduce them. If a student performs poorly on a test, compare his or her exam to its corresponding Test-Item Analysis, placing a check mark beside every problem answered incorrectly.

Next compare the checked items. Are they on related topics? If so, the student might need to be retaught those topics. (See "Reteaching" below.) Are the lessons for the checked items clustered? If so, the student might need more practice on that section of lessons. (See "Additional Practice" and "Recycling" below.)

Analysis of Whole-Group Performance

Recording Form C (Test Analysis) allows you to analyze an entire class's performance on a single exam. It provides a framework to tally the number of students who incorrectly answered each test problem. Analyzing this data will help you determine which concepts, if any, need to be retaught to the whole class.

Student Self-Analysis

Recording Form D (Individual Test Analysis) allows students who score below 80 percent on a test to analyze their own work. It asks them to identify whether their mistakes are conceptual or simply careless. After a student completes the self-analysis, discuss it with him or her. Plan remediation based on the "Reteaching," "Additional Practice," and "Recycling" sections below. Write your agreed-upon course of action at the bottom of the form.

Reteaching

One reason a student might not understand a concept after repeated practice could be that the teaching of the concept was incompatible with the student's learning style. Perhaps the student is a tactile learner, but the concept was presented visually, or maybe the student is a visual learner, but the instruction was mostly oral. Awareness of learning styles is critical when reteaching. As much as possible, tailor your instruction to the individual. If manipulatives were not used the first time or if visuals were not presented, incorporate them when you reteach. The Teacher's Manual often has specific ideas.

Another pitfall is overly complex instruction. The hints in this guide are an excellent resource to help you avoid complicated instruction. Many of them reduce the complex to the simple to help struggling students. Other hints provide alternate instruction to that in the Student Edition or Teacher's Manual, giving options for teaching certain lessons.

Sometimes the best reteaching begins with asking the student to explain the concept. Often his or her misunderstandings will be evident, and your reteaching can address the revealed points of confusion.

Additional Practice

Supplemental Practice

Supplemental Practices provide additional practice for the more challenging lessons.

Although the *Saxon Math 5/4* textbook contains sufficient practice for most students to learn the concepts, some students might require additional practice to fully grasp certain topics.

The Student Edition contains Supplemental Practice exercises for select topics that are typically more difficult to understand. You can ask students to complete the exercises on notebook paper or to use the Supplemental Practice Worksheets from the Student Workbook. (For maximal support, opt for the worksheets, which explain how to solve the problems and often show examples.) The chart below identifies the topic of each Supplemental Practice in *Saxon Math 5/4* and the lesson it supports.

USE WITH LESSON	TOPIC
16	Two-Digit Subtraction with Regrouping/Missing Numbers in Subtraction
17	Column Addition
30	Three-Digit Subtraction
34	Writing and Reading Numbers
37	Fractions and Mixed Numbers on a Number Line
41	Subtracting Across Zero
43	Adding and Subtracting Dollars and Cents
48	Multiplying Two-Digit Numbers
50	Adding and Subtracting Decimal Numbers
52	Subtracting Numbers with More Than Three Digits
53	One-Digit Division with a Remainder
58	Multiplying Three-Digit Numbers
64	Division with Two-Digit Quotients
65	Division with Two-Digit Quotients
67	Multiplying by Multiples of 10
68	Division with Two-Digit Quotients and a Remainder

USE WITH LESSON	TOPIC
76	Division with Three-Digit Quotients
80	Division with Zeros in Three-Digit Quotients
90	Multiplying Two Two-Digit Numbers
104	Converting Improper Fractions
107	Adding and Subtracting Mixed Numbers and Fractions with Common Denominators
110	Dividing by Multiples of 10
112	Reducing Fractions
113	Multiplying a Three-Digit Number by a Two-Digit Number
114	Simplifying Fraction Answers
115	Equivalent Fractions
118	Dividing by Two Digits
119	Adding and Subtracting Fractions with Different Denominators

Fraction Activities

Fraction Activities aid the transition from tactile to pencil-and-paper tasks involving fractions.

The Fraction Activity Masters (found in the *Intervention* Masters) are one-page worksheets designed to be completed with the use of manipulatives. They provide extra practice for students who struggle with fraction concepts. Guidelines for usage are given in the *Intervention* Masters. The chart below identifies the topic of each Fraction Activity Master and the lesson it supports.

FRACTION ACTIVITY	USE WITH LESSON	TOPIC
A	22	Naming Fractions of Shapes
B	26	Drawing Pictures of Fractions
C	Inv. 5	Comparing Fractions, Decimals, and Percents
D	56	Comparing Fractions
E	61	Remaining Fractions
F	89	Mixed Numbers and Improper Fractions
G	95	Solving Fraction-of-a-Group Problems
H	107	Adding and Subtracting Like Fractions
I	109	Equivalent Fractions
J	112	Simplifying Mixed Numbers

Although the *Saxon Math 5/4* Test and Practice Generator (a CD-ROM for optional purchase) is primarily intended to allow teachers to customize and randomize their exams, it can also be used to create concept-specific practice sheets for any topic in the textbook. Please note, however, that this capability should be used sparingly, not as a routine supplement to the daily assignments. Besides making the assignments too long, massed practice is not required for students to learn most concepts. The distributed mixed review built into the program allows students numerous opportunities to practice and master concepts and provides teachers numerous opportunities to clarify, reteach, and reinforce. Finally, never replace any portion of the daily assignment with concept-specific practice. The **distributed mixed review is vital** to students' long-term understanding of mathematics.

Recycling

Recycling is best suited for classrooms using self-paced instruction.

If a student *can* do the work, he or she normally *will* do the work. Therefore, if a student begins to demonstrate negative behavior or passive resistance, it might be because the student is being asked to do something that is too difficult for him or her at that time. If this happens frequently, the student might experience frustration and a sense of failure in math. This student needs another chance—a chance to recycle concepts not mastered.

Recycling is a process in which students repeat a series of lesson practice sets and/or problem sets. It is recommended for students who experience difficulty with concepts on the tests. Recycling does not hurt the student's grade—it usually helps the grade because the student will likely grasp missed concepts the second time through. Because students already have some familiarity with the lesson practice sets, the problems can be covered again more rapidly.

More than five conceptual errors on a test usually indicates a need for recycling. (See the testing schedule on page 9 to determine which lessons to repeat.) However, use your judgment, because errors might be due more to carelessness than to a lack of understanding of concepts. In this case, offer students an opportunity to correct errors independently. This will be the most effective way to judge concept mastery.

Recycling is also suggested when a student performs poorly on subsequent assignments. If a student misses ten or more problems of a Mixed Practice on the first attempt, something might be wrong. Note the first time this occurs; if it happens again within five days, the student should recycle the five lessons preceding the first occurrence.

Recycling lessons is very beneficial to a struggling student. The student might be slightly embarrassed but will ultimately be greatly relieved to have a chance to become better prepared before moving on to more difficult work. If a student refuses to recycle, give him or her the time and space to calm down and think about the benefits of recycling:

- The material will be familiar, which will lead to better scores.
- Assignments will probably take less time to complete.
- There will probably be fewer errors to correct.

Assigning a bonus point for recycling should help with both attitude and grading.

Decreasing Usage of Support Materials

If allowed to choose, students will often discontinue their use of worksheets at the appropriate time.

Some students will be able to transition away from using *Intervention* materials partway through the school year. Use the following practices to facilitate the change:

- Throughout the year, make worksheet use optional. Many students will elect to move away from them on their own. (Note: You might have to monitor this as a few students will attempt the transition too soon.)
- Do not allow capable students to use the worksheets as a crutch.
- Gradually reduce students' use of worksheets. Let students try one lesson without the corresponding worksheet. Students who score 80 percent or better can complete the next lesson the same way. Students who score below 80 percent can try again in a couple of weeks.
- Remind students throughout the year that they will not be able to use reference guides on end-of-year, standardized exams. In early spring, begin impressing upon students the need to cut back their use of reference guides on tests. Most students will do so on their own.
- Copy the Answer Forms from the *Intervention* Masters for students to use as they transition from worksheets to notebook paper. Answer Forms establish a workspace for each problem but do not contain the helps found on worksheets.

Not all students should transition away from using *Intervention* materials. Continue to use the materials with students who consistently function below their target grade level.

Placement

The Middle Grades Placement Test is available online at www.saxonpublishers. com. The student's score will be computed automatically.

To place a student in the proper grade level, begin by looking at:

- past performance in math classes
- scores on standardized tests
- recommendations from previous teachers

If the student has not used Saxon Math™ lately, and if the available information is inadequate for you to confidently select the textbook a student will use, administer the Middle Grades Placement Test to gain further insight. The test is available online at www.saxonpublishers.com; you can also request one in print form by calling (800) 284-7019 and asking for a Middle Grades Teacher's Resource Booklet. Scores are computed automatically for the online version; an answer key is provided with the print version.

The placement test can be used to roughly determine whether a student should start in *Saxon Math 5/4, 6/5, 7/6,* or *8/7.* Use the following guidelines when administering the test to students with special needs. (Note that these differ slightly from the instructions provided with the test itself.)

- Do not time the test.
- The student should show all work and should work through the test until he or she cannot work any more problems.
- Reading assistance may be provided for low-level readers so that the assessment of their math skills will not be affected by their reading ability.

- The student must not use a calculator.
- Use the chart below to guide placement.

Middle Grades Placement Test Guide

Four or fewer correct out of Questions 1–10	Student may begin *Saxon Math 5/4*.
Five or more correct out of Questions 1–10	Student may begin *Saxon Math 6/5*.
Five or more correct out of Questions 11–20	Student may begin *Saxon Math 7/6*.
Five or more correct out of Questions 21–30	Student may begin *Saxon Math 8/7*.
Five or more correct out of Questions 31–40	Student may begin *Saxon Algebra $\frac{1}{2}$*.

Once you determine the grade level, you can identify the proper starting point within that level (if you use self-paced instruction). Begin by administering Test 1, and continue through the tests until four or more **conceptual** errors are made on a single test. Then find the appropriate starting lesson in the chart below. (You can begin the process with a later test if you have some awareness of the student's abilities. Just be sure not to start too high. If the student makes six or more conceptual errors on the first test, you should restart the process and begin with an earlier test.)

Placement Within a Textbook

TEST NUMBER	PLACEMENT WITH 4 OR 5 CONCEPTUAL ERRORS	PLACEMENT WITH 6 OR MORE CONCEPTUAL ERRORS
1	Lesson 6	Lesson 1
2	Lesson 11	Lesson 6
3	Lesson 16	Lesson 11
4	Lesson 21	Lesson 16
5	Lesson 26	Lesson 21
6	Lesson 31	Lesson 26
7	Lesson 36	Lesson 31
8	Lesson 41	Lesson 36
9	Lesson 46	Lesson 41
10	Lesson 51	Lesson 46
11	Lesson 56	Lesson 51

Placement Within a Textbook (continued)

TEST NUMBER	PLACEMENT WITH 4 OR 5 CONCEPTUAL ERRORS	PLACEMENT WITH 6 OR MORE CONCEPTUAL ERRORS
12	Lesson 61	Lesson 56
13	Lesson 66	Lesson 61
14	Lesson 71	Lesson 66
15	Lesson 76	Lesson 71
16	Lesson 81	Lesson 76
17	Lesson 86	Lesson 81
18	Lesson 91	Lesson 86
19	Lesson 96	Lesson 91
20	Lesson 101	Lesson 96
21	Lesson 106	Lesson 101
22	Lesson 111	Lesson 106
23	Lesson 116	Lesson 111

Placement is correct if the first few lessons are very easy for the student to complete. Students are more likely to be successful if they are allowed to develop appropriate classroom behavior and good homework habits without the additional challenge of difficult assignments.

Individualized Education Plans

A list of short-term Individualized Education Plan (IEP) objectives for *Saxon Math 5/4* is provided on pages 62–63. Each objective is followed by the lesson number after which the concept is tested (based on testing schedule B from page 9). Because each concept will be practiced at least ten times before it is tested, most students achieve 80 percent accuracy. With this list, teachers can determine which objectives have already been met and which should apply to the coming year.

These short-term instructional objectives are in a ready-to-use format. Teachers may copy and complete a set of the pages for each student or just use the list to select certain objectives for each student's IEP. Teachers may also input these objectives on a computer, make a copy of the file for each student, and delete the portions that do not apply to a particular student. A copy of these objectives can be downloaded from the Saxon Web site (www.saxonpublishers.com) and modified as necessary.

Short-Term Instructional Objectives for *Saxon Math 5/4*

Instructional Objectives: **Date Projected:**   **Evaluation Results:**

- Given a series of written computation and word problems involving **whole numbers,** the student will complete them with 80 percent accuracy as measured by teacher-made tests.
    1. Adding to three digits with regrouping (23)[†]   _____   _____
    2. Subtracting to three digits with regrouping (40)   _____   _____
    3. Reading and writing place value through thousands (44)   _____   _____
    4. Multiplying two digits by one digit with regrouping (58)   _____   _____
    5. Dividing three digits by one digit (86)   _____   _____

- Given a series of written computation and word problems involving **fractions,** the student will complete them with 80 percent accuracy as measured by teacher-made tests.
    1. Recognizing fraction value (32)   _____   _____
    2. Comparing fractions using pictures (66)   _____   _____
    3. Recognizing fraction of a group (80)   _____   _____
    4. Adding and subtracting with like denominators (117)   _____   _____
    5. Reducing and converting improper fractions to mixed or whole numbers (120)   _____   _____

- Given a series of written computation and word problems involving **decimals and percents,** the student will complete them with 80 percent accuracy as measured by teacher-made tests.
    1. Changing decimals to fractions (51)   _____   _____
    2. Adding and subtracting decimals to hundredths place by aligning place value (53)   _____   _____
    3. Changing fractions to percents (61)   _____   _____
    4. Finding the percent of a whole using pictures (61)   _____   _____

- Given a series of written problems involving **number theory,** the student will demonstrate understanding by working them with 80 percent accuracy as measured by teacher-made tests.
    1. Comparing whole numbers (21)   _____   _____
    2. Sequences and patterns in the multiplication table (38)   _____   _____
    3. Listing multiples of whole numbers (65)   _____   _____
    4. Listing factors of whole numbers (65)   _____   _____

- Given a series of written computation and word problems involving **money concepts,** the student will complete them with 80 percent accuracy as measured by teacher-made tests.
    1. Writing dollars and cents in correct form (45)   _____   _____
    2. Adding and subtracting money (53)   _____   _____
    3. Multiplying and dividing money (86)   _____   _____
    4. Sales tax (93)   _____   _____

- Given a series of written problems involving the **interpretation of graphs and tables,** the student will complete them with 80 percent accuracy as measured by teacher-made tests.
    1. Pictographs, bar graphs, line graphs, and circle graphs (71)   _____   _____
    2. Tables (111)   _____   _____

[†]The number in parentheses refers to the lesson number after which the concept may be tested according to testing schedule B.

Instructional Objectives (continued):   Date Projected: Evaluation Results:

- Given a series of written problems involving **geometry,** the student will complete them with 80 percent accuracy as measured by teacher-made tests.

    1. Calculating perimeter of polygons (31)   _____   _____
    2. Identifying lines, segments, and rays (33)   _____   _____
    3. Recognizing right, acute, and obtuse angles (33) _____   _____
    4. Calculating area of rectangles (41)   _____   _____
    5. Identifying polygons by number of sides (73)   _____   _____
    6. Recognizing geometric solids (108)   _____   _____

- Given a series of written computation and word problems involving **measurement,** the student will complete them with 80 percent accuracy as measured by teacher-made tests.

    1. Reading scales (28)   _____   _____
    2. Abbreviations of metric and U.S. Customary units of length (31) _____   _____
    3. Elapsed time (37)   _____   _____
    4. U.S. Customary System of length, liquid, and weight (87) _____   _____
    5. Metric system of length, liquid, and weight (87) _____   _____

- Given a series of written problems involving **algebraic concepts,** the student will complete them with 80 percent accuracy as measured by teacher-made tests.

    1. Finding the unknown in equations (51)   _____   _____
    2. Using parentheses as symbols of inclusion (55) _____   _____
    3. Solving two-step equations (71)   _____   _____

Sample IEP Form

School District
Special Education Local Plan Area
Instructional Plan—Goals and Objectives

Student's Name **Tanesha Doe**   IEP Date **8/16/04**

☑ RS   ☐ SC   ☐ Home/Hospital

Annual Goal(s)

Tanesha will improve in calculation and math applications.

Present Level of Performance

She is in the early stages of multiplying fractions.

Instructional Objectives	Date Projected	Evaluation Results
■ Given a series of written computation and word problems involving **fractions,** the student will complete them with 80 percent accuracy.		
1. Recognizing fraction value (32)		
2. Comparing fractions using pictures (66)		
3. Recognizing fraction of a group (80)		
4. Adding and subtracting with like denominators (117)		
5. Reducing and converting improper fractions to mixed or whole numbers (120)		
■ Given a series of written computation and word problems involving **decimals and percents,** the student will complete them with 80 percent accuracy.		
1. Changing decimals to fractions (51)		
2. Adding and subtracting decimals to hundredths place by aligning place value (53)		
3. Changing fractions to percents (61)		
4. Finding the percent of a whole using pictures (61)		
■ Given a series of written problems involving **number theory,** the student will complete them with 80 percent accuracy.		
1. Comparing whole numbers (21)		
2. Sequences and patterns in the multiplication table (38)		
3. Listing multiples of whole numbers (65)		
4. Listing factors of whole numbers (65)		

Method of Evaluation   ☐ Observation   ☐ Criterion-Referenced Tests   ☑ Teacher-Made Tests
☐ Standardized Tests   ☐ Other _____

Needed Specialized Equipment and Services

Summary of Progress *(to be completed at next IEP meeting)*

Suggested Reading

Armstrong, Thomas. *In Their Own Way: Discovering and Encouraging Your Child's Multiple Intelligences.* Los Angeles: J. P. Tarcher, 2000.

Hallahan, Daniel P., James M. Kauffman, and John W. Lloyd. *Introduction to Learning Disabilities* (2nd Edition). Boston: Pearson Allyn & Bacon, 1998.

Harwell, Joan M. *Complete Learning Disabilities Handbook: Ready-to-Use Strategies and Activities for Teaching Students with Learning Disabilities,* 2nd ed. New York: Jossey-Bass, 2002.

Osman, Betty B. *Learning Disabilities and ADHD: A Family Guide to Living and Learning Together.* New York: John Wiley & Sons, 1997.

Parker, Harvey C. *The ADD Hyperactivity Handbook for Schools: Effective Strategies for Identifying and Teaching Students with Attention Deficit Disorders in Elementary and Secondary Schools.* Plantation, FL: Impact Publications, 1992.

Platt, Jennifer M., and Judy L. Olson. *Teaching Adolescents with Mild Disabilities.* Pacific Grove, CA: Brooks/Cole Publishers, 1996.

Silver, Larry B. *The Misunderstood Child: Understanding and Coping with Your Child's Learning Disabilities.* New York: Random House, 1998.

Smith, Corinne, and Lisa Strick. *Learning Disabilities: A to Z. A Parent's Complete Guide to Learning Disabilities from Preschool to Adulthood.* New York: Simon & Schuster, 1999.

Smith, Sally L. *No Easy Answers: The Learning Disabled Child at Home and at School.* New York: Bantam Books, 1995.

Stevens, Suzanne H. *The LD Child and ADHD Child: Ways Parents and Professionals Can Help.* Winston-Salem, NC: John F. Blair Publications, 1996.

Wender, Paul H. *The Hyperactive Child, Adolescent, and Adult: Attention Deficit Disorder Through the Lifespan.* New York: Oxford University Press, 1987.

Young, Rosalie, and Harriet H. Savage. *How to Help Students Overcome Learning Problems and Learning Disabilities: Better Learning for All Ages,* 2nd ed. Danville, IL: Interstate Printers and Publishers, 1989.

Suggested Viewing

How Difficult Can This Be? A Learning Disabilities Workshop (F.A.T. City) (1989)
Seventy-minute video with discussion guide. Order online: http://ldonline.org/ld_store/ld.html. Order by phone: (800) 343-5540.

Learning Disabilities and Discipline with Richard Lavoie: When the Chips Are Down … Strategies for Improving Children's Behavior (1997)
Sixty-two-minute video and discussion guide. Order online: http://ldonline.org/ld_store/ld.html. Order by phone: (800) 343-5540.

Learning Disabilities and Social Skills with Richard Lavoie: Last One Picked … First One Picked On (1994)
Sixty-eight-minute teacher version video with teacher's guide or sixty-two-minute parent version video with parent's guide. Order online: http://ldonline.org/ld_store/ld.html. Order by phone: (800) 343-5540.

Related Organizations

Children and Adults with Attention-Deficit/Hyperactivity Disorder (CHADD)
8181 Professional Place, Suite 150
Landover, MD 20785
(800) 233-4050 or (301) 306-7070
www.chadd.org

The Council for Exceptional Children (CEC)
1110 North Glebe Road, Suite 300
Arlington, VA 22201-5704
(888) CEC-SPED or (703) 620-3660
www.cec.sped.org

The International Dyslexia Association (IDA)
(formerly Orton Dyslexia Society)
Chester Building, Suite 382
8600 LaSalle Road
Baltimore, MD 21286-2044
(800) ABCD123 or (410) 296-0232
www.interdys.org

Learning Disabilities Association of America (LDA)
4156 Library Road
Pittsburgh, PA 15234-1349
(412) 341-1515
www.ldanatl.org

National Center for Learning Disabilities (NCLD)
381 Park Avenue South, Suite 1401
New York, NY 10016
(888) 575-7373 or (212) 545-7510
www.ncld.org